PEATLANDS

This book provides an introduction to peatlands for the non-specialist student reader and for all those concerned about environmental protection, and is an essential guide to peatland history and heritage for scientists and enthusiasts.

Peat is formed when vegetation partially decays in a waterlogged environment and occurs extensively throughout both temperate and tropical regions. Interest in peatlands is currently high due to the degradation of global peatlands which is disrupting hydrology and contributing to greenhouse gas emissions. This book opens by explaining how peat is formed, its properties and worldwide distribution, and defines related terms such as mires, wetlands, bogs and marshes. There is discussion of the ecology and wildlife of peatlands as well as their ability to preserve pollen and organic remains as environmental archives. It also addresses the history, heritage and cultural exploitation of peat, extending back to pre-Roman times, and the degradation of peatlands over the centuries, particularly as a source of fuel but more recently for commercial horticulture. Other chapters discuss the ecosystem services delivered by peatlands, and how their destruction is contributing to biodiversity loss, flooding or drought, and climate change. Finally, the many current peatland restoration projects around the world are highlighted.

Overall the book provides a wide-ranging but concise overview of peatlands from both a natural and social science perspective, and will be invaluable for students of ecology, geography, environmental studies and history.

Ian D. Rotherham is a Professor of Environmental Geography, Reader in Tourism and Environmental Change and International Research Coordinator at Sheffield Hallam University, UK. For several years he was the chair and secretary of the British Ecological Society's Peatlands Research Special Interest Group. He is the author or editor of numerous books on topics in ecology, environmental studies and environmental history. These include: *The Rise and Fall of Countryside Management* (Routledge, 2016); *Trees, Forested Landscapes and Grazing Animals* (Routledge, 2013); and *Invasive and Introduced Plants and Animals* (Routledge, 2011).

Earthscan Studies in Natural Resource Management

Social Movements Contesting Natural Resource Development
Edited by John F. Devlin

Benefit-sharing in Environmental Governance
Local Experiences of a Global Concept
Louisa Parks

Governing Renewable Natural Resources
Theories and Frameworks
Edited by Fiona Nunan

Interdisciplinary Collaboration for Water Diplomacy
A Principled and Pragmatic Approach
Edited by Shafiqul Islam and Kevin M. Smith

Peatlands
Ecology, Conservation and Heritage
Ian D. Rotherham

The Moroccan Argan Trade
Producer Networks and Human Bio-Geographies
Daniel F. Robinson

For more information on books in this series, please visit: http://www.routledge.com/books/series/ECNRM/

PEATLANDS

Ecology, Conservation and Heritage

Ian D. Rotherham

Routledge
Taylor & Francis Group
LONDON AND NEW YORK

earthscan
from Routledge

First published 2020
by Routledge
2 Park Square, Milton Park, Abingdon, Oxon OX14 4RN

and by Routledge
52 Vanderbilt Avenue, New York, NY 10017

Routledge is an imprint of the Taylor & Francis Group, an informa business

British Library Cataloguing-in-Publication Data
A catalogue record for this book is available from the British Library

Library of Congress Cataloguing-in-Publication Data
Names: Rotherham, Ian D., author.
Title: Peatlands: ecology, conservation and heritage / Ian D. Rotherham.
Description: Abingdon, Oxon; New York, NY: Routledge, 2020. |
Series: Earthscan studies in natural resource management |
Includes bibliographical references and index.
Identifiers: LCCN 2019056956 (print) | LCCN 2019056957 (ebook) |
ISBN 9781138343207 (hardback) | ISBN 9781138343214 (paperback) |
ISBN 9780429439285 (ebook)
Subjects: LCSH: Peatlands. | Peatland ecology. | Peatland restoration. |
Peatland management.
Classification: LCC GB621 .R67 2020 (print) | LCC GB621 (ebook) |
DDC 577.68/7—dc23
LC record available at https://lccn.loc.gov/2019056956
LC ebook record available at https://lccn.loc.gov/2019056957

ISBN: 978-1-138-34320-7 (hbk)
ISBN: 978-1-138-34321-4 (pbk)
ISBN: 978-0-429-43928-5 (ebk)

Typeset in Bembo
by codeMantra

'This volume is dedicated to the late Professor David Bellamy; the man who inspired so many of us to battle for the peat-bogs and whose presence will be missed for decades to come.'

CONTENTS

FOREWORD

Prince Charles, for example, has likened Britain's peatlands to the environmental equivalent of tropical rainforests. Sadly, like the rainforests, the world's peatlands are also in rapid decline, and again like the forests, the implications for people and for 'ecosystem services' are very bad. With this as the context, the need to understand peat and peatlands has grown to parallel their loss. The intention of this book is to help bridge the gap of engaging and informing professionals such as ecologists and land managers about this 'back gold' under our feet and maybe extending the subject to a wider audience too.

This introductory volume fulfils a niche in the literature on peat and peatlands in covering aspects of ecology, conservation management, and heritage. With many on-going restoration projects, interest in peat and peatlands is at an all-time high. However, the literature available on peatlands is still quite limited and furthermore, focused almost entirely on aspects of hydrology, restoration, and the ecology associated with these.

In this context, the book considers the remarkable material known as 'peat', its particular properties, and how it is formed. The occurrence and nature of peat and peatlands are examined and explained along with key aspects of its hydrology and other physical properties. The varying types of peat and peatlands which occur around the world are introduced and discussed. There is reference to the ecology and wildlife of these remarkable landscapes, and the impacts of drainage and other damaging operations. The importance of peatlands as environmental archives for pollen and insect remains and for other preserved materials is considerable and provides insight into environmental changes over thousands of years. This is rarely recognised in conservation or indeed in 'restoration' projects since it is a finite resource which once destroyed cannot be recovered or re-created.

Having established the nature, occurrence, and importance of peatlands globally, and briefly noted the problems facing many sites, the book examines the history and cultural exploitation of peat, and the consequences of such human impacts extending back to pre-Roman times. This is followed by a discussion of dramatic human impacts on peatlands through the period from the 1700s to late 2000s, and the consequences of the massive removal and widespread degradation. The implications for biodiversity loss are generally obvious, but peat removal has also triggered major breakdowns in ecosystem service delivery with consequent problems of flooding, drought, and perhaps even climate change. Furthermore, peatland degradation and removal have at the very least exacerbated climate shifts associated with carbon release. However, a key impact of the changes is that they have also reduced the ecosystem capacity to buffer any consequent changes in environmental stresses such as climate warming. The absence of extensive peatlands in good condition compromises the ability of the landscape to mitigate climatic fluxes and to ameliorate the adverse impacts of such changes.

Today UK peatlands damaged by problems such as fires and droughts may release around 381,000 tonnes of carbon back to the atmosphere each year. The problems are not evenly distributed as the conditions of the remaining bogs vary dramatically with the location. So by the early twenty-first century, for example, it became clear that whilst peat bogs in say Scotland were still absorbing atmospheric carbon, those sites elsewhere in the UK were not. In Wales and England, for example, because of gross air pollution impacts and aggressively intensive drainage and farming, the bogs and other peatlands are not only failing to mop up atmospheric carbon but are actively releasing it back to the air and peat-stained waters back into water supplies. The costs, direct and indirect, are immense.

The need is now to set in place the circumstances through which the peatlands can be rejuvenated, though this will be neither easy nor quick. However, the book closes with an overview of positive restoration projects and approaches and the future for peatlands. A more intransigent challenge is perhaps that of the peatlands converted to intensive arable farming. Many, like the English fenlands for example, have only a few decades of soil utilisation left. All the while that these lands are being cultivated they are in effect haemorrhaging carbon back into the atmosphere as carbon dioxide. So despite the good works being done to restore damaged peatlands, particularly in degraded blanket bog areas and in some lowland raised mires, there is still much more to do. The lowland 'bread-basket' fenland areas in particular are at risk from soil erosion, drought, and coastal inundation and 'saline creep' (saltwater moving into the groundwater aquifers), and the long-term implications for food security, for climate change, and more are hugely worrying.

FIGURE 0.2 Traditional peat-cutting at the turbary, in Ireland in the 1970s.

Much of the book is focused on the British Isles because that is where much of my personal interest lies. However, the wider context is considered with examples from around the world, and the global situations are introduced and discussed. Furthermore, peatlands are truly global in extent, and their catastrophic misuse over the last century, in particular, is a worldwide issue with all-encompassing implications for people, biodiversity, climate, and all our futures.

PREAMBLE

FIGURE 0.3 Sphagnum carpet on Ringinglow Bog in the English Peak District.

Peat is a precious resource and peatlands are unique and special habitats. Peatlands are remarkable places to visit and alongside rare and threatened species, they have rich cultural and historical associations and economic functions too. Indeed, peatlands have been a part of the British and European cultural heritage since the Iron Age at least, that is, around 2,000 years. In past times, some communities, especially along the western Atlantic seaboard, relied almost entirely on peat as their source of fuel. Today most people's connection to peat is through its widespread use in horticulture, and they have little idea of its rich biodiversity or its depth of cultural heritage. I have suggested that peat bogs are 'cathedrals of biodiversity' and yet, in recent decades around the world, we have destroyed them almost relentlessly and we continue to do so. Some of this destruction is merely to access the landscape for other purposes such as the drainage and burning today of Indonesia's tropical peat forests for so-called 'sustainable' palm-oil plantations. In this context, to describe the usage as 'sustainable' beggars belief. Other damage wrought over centuries has been the incremental harvesting of peat as fuel for domestic household use and more recently for industry. In Ireland, for example, vast quantities of peat-fuel were cut industrially in the twentieth century by the state-owned energy company Bord Na Móna. Even today, despite publicity to the contrary, Ireland continues to export vast amounts of peat for horticultural compost and soil conditioner. This is the ecological equivalent of demolishing York Minster or Canterbury Cathedral and using the stone to make rockeries for domestic gardens.

However, peatlands are not merely wonderful habitats for a rich, diverse, and often unique wildlife. The types of peatlands and associated habitats such as heather moorland vary markedly with differing environmental conditions. The peat itself can be formed in different ways from various types of vegetation under different conditions; fen peats from reed and sedge, for example, can be quite base-rich and differ significantly from bog peats such as from acidic upland blanket mires. These variations have a big impact on the functioning of the various peatlands and their associated fauna and flora. The UK has around 75% of the global resource of heather moorland and much of this is a peat-based habitat. Also in upland or northern environments, the UK has about 10% to 15% of the global areas of blanket mire or blanket bog. We formerly had extensive lowland fens and also raised mires, but these have been massively reduced in recent centuries and the rate of loss accelerated throughout the twentieth century.

Peatlands also have huge value in delivering so-called 'ecosystem services' such as water management, carbon storage, and climate-change mitigation. In the UK alone, the remaining peatlands store around three billion tonnes of atmospheric carbon and those which remain as functioning ecosystems are actively mopping up more carbon from the air by a process called sequestration. The amount currently stored is more than the equivalent of that stored in all the woods and forests of France and Great Britain combined. Recent calculations have suggested that pristine peatlands in the UK absorb around 400,000 tonnes of carbon each year. In tropical areas which are hot and wet, the process of absorption is even faster. Yet despite recognition of these valuable services, peatlands globally are still massively over-exploited and this is especially the case in tropical regions. The destruction of peatlands historically in temperate zones, and today in the tropics, must rate as one of the worst environmental catastrophes of all time and yet goes on with little popular recognition. Even key decision-makers and politicians have demonstrated little understanding of what is happening and of its implications.

FIGURE 0.4 Surveying for sphagnum mosses in the English Peak District.

Nevertheless, by the late twentieth century, policies across Britain and Europe had at last begun to slow the exploitation of regional peat resources. In recent years, projects to begin the slow process of recovery have begun. However, whether there will be any real slowing of the destruction of globally important tropical resources we cannot predict. Long-term environmental, social, and economic consequences are invariably pushed aside by short-term interests of immediate financial gain through exploitation and destruction.

ACKNOWLEDGEMENTS

I wish to acknowledge the many friends and colleagues who are peatland enthusiasts and have helped and encouraged over many years. I also thank the editorial team at Routledge for their support and assistance, and in particular for their patience.

GLOSSARY OF TERMS

Acrotelm the upper layer of a raised bog above the saturated zone of the 'catotelm'

Active porosity the ability of water to move within the peat body of particles and inter-particle water

Anoxic lacking oxygen or anaerobic

Aquifer water-bearing strata

Baulks upstanding blocks of peat or other materials between peat-cut fields

Block-cutting method of peat-cutting during which bocks are cut from a surface face of peat

Bog or bogs a generic term of ombrotrophic mires but often applied colloquially to other wetland systems

Bog species fauna or flora typically associated with bogs

Bog-moss commonly used term for the 'sphagnum mosses'

Bog vegetation plant communities typically associated with bogs

Bulk density the amount of solid material per unit volume

Bund an embankment generally used to hold back water – may be synonymous with a dam, embankment, or dyke

Catotelm the lower, saturated, and relatively inert layer of peat in an undamaged raised bog

Cut-over surface a generic term for the peat (or other) surface left after peat-cutting or harvesting

Draw-down fall in water level as caused by hydraulic gradient to say a ditch, drain, or dyke

Eutrophic of soil or water, nutrient-enriched

Evapotranspiration loss of water from a vegetation or soil surface by a mix of evaporation from the soil and transpiration from plants

Fen generic term for minerotrophic mire

Flats the base of peat-cuttings in a block-cut system
Flooding in a peatland refers to inundation of the peat surface
Gley soil formed under conditions of low or fluctuating oxygen levels
Humification degree of decomposition of peat or the production of humus in soil due to breakdown of organic matter
Hummock surface of a peat bog raised perhaps 20 cm to 50 cm above the lower surface level and consequently drier and characterised by, for example, ericaceous plants and lichens
Hydraulic gradient change in hydraulic head or water surface elevation over a distance
Hydraulic head difference in hydraulic pressure between two connected points
Hydrosere ecological succession leading to terrestrialisation of open water
Hydrostatic pressure pressure generated by the weight of water
Iron pan hard and impermeable layer formed in a podzol soil horizon under conditions of low or fluctuating oxygen levels and fluctuating water levels
Lagg the outer zone of a raised bog
Lawn peatland vegetation dominated by grasses and sedges with a range of bryophyte species and generally between 5 cm and 20 cm above the lower layer
Macrofossils plant and animal remains found in peat and capable of being identified merely by means of a microscope
Marsh this is wetland vegetation which is mostly found on minerals soils but on peatland areas too. It generally is formed where there is open water or flowing water and is at least seasonally inundated
Meteoric water water from rainfall, that is, precipitation
Milling modern (twentieth-century) method of peat extraction that uses surface milling, harrowing, and drying before collection of the worked peat
Mineralisation oxidative breakdown of organic matter into its constituent inorganic components and carbon dioxide
Minerotrophic fed by groundwater
Mire wetland with at least some peat accumulation and dominated by peat-forming vegetation
Oligotrophic of soil or water, nutrient-poor
Ombrotrophic supplied with water derived solely from the atmosphere, that is rain, snow, fog, mist, etc.
Ombrotrophic mire or bog bog or mire solely gaining water via atmospheric input
Palaeo-ecology study of past organisms and their environment
Paludification development of wetland such as a bog directly over mineral soil or rock due to waterlogging, increased water input, and impeded drainage
Peat undecomposed or part decomposed remains of plant and animal material which collects under conditions which are water-logged and anaerobic
Peat-cuttings areas from which peat has been removed by cutting rather than milling

Peat-field areas between drains in an open milled peatland
Peatland in the strict sense a peatland is terrain covered with peat to a depth of at least 30 cm to 40 cm though this varies between definitions
Peat-workings generic terms for a site worked for peat extraction by any method
Perched water mound mound developed in a raised bog due to impeded drainage
Permeability ability of a porous medium to transmit water
Pool permanently water-filled basin in a peatland and a feature which is initiated and deepened as the mire develops
Poor fen generally minerotrophic mire but with pH about 5.5
Porosity or pore space portion of the soil medium occupied by air and water
Precipitation atmospheric fallout of water via rain, snow, etc.
Rafting growth of plants and vegetation floating on water or in a 'supra-water' layer
Rand outward-sloping margins of a raised bog
Recharge water input or supply to the system
Regeneration new growth of peat as a mire recovers from exploitation
Remnant usually peat-blocks left on-site after cutting
Re-naturation recovery of ombrotrophic vegetation following disruption or damage
Rewetting restoration or recovery of wet conditions on the mire surface
Rich fen minerotrophic mire generally with pH above 5.5
Runoff water exiting a mire system due to gravity drainage
Schwingmoor German word for a floating vegetation mat
Sedimentary deposits deposits in ponds, lake, the sea, or other water-bodies with material washed in, deposited from the atmosphere, or generated within the water system and accumulating as sediment
Sere vegetation successional stage
Soligenous minerotrophic peatlands on a slope and with water flowing directionally through the peat or across the surface
String elongated ridges in patterned bogs and fens which are organised perpendicularly to the slope and with a hummock or lawn vegetation
Subsidence sinking of a peat surface after drought or drainage and due to collapse of pore-spaces, loss of water, and aerobic decomposition
Swamp or swamp-forest forested wetland on either minerotrophic soil or minerotrophic peat but separated from wooded fen or carr by virtue of a denser tree canopy (greater than 25% cover)
Terrestrialisation (or infilling) the successional process whereby vegetation at the margins of open water and fen develops into the centre of a pond, lake, or riverine margin
Topogenous minerotrophic peatland with near horizontal water table and located in a basin
Waterlogging the water table situated close to the surface of the soil

Water storage volume of water held in a given amount of soil or peat, for example

Water storage capacity total amount of water which can be held by a given volume of soil

Water table this is the level at which pore-water pressure equals atmospheric pressure and is the junction between saturated and unsaturated conditions – the upper surface of the zone of saturation which is where pores and fractures in the substrate are saturated with water. It is also the level to which water will rise up in a hole within the peat mass and is the upper surface of the groundwater

Wetland land where the water table is close to the surface and inundation persists for a large part of the year so that organisms must be adapted to wet and anoxic conditions

PART 1

Setting the scene

1

INTRODUCTION AND OVERVIEW

Peat and peatlands

An introduction to peat and peat-cutting

The Roman writer Pliny in his *Natural History* described peat-cutting in the first century AD, by German tribes along the Rivers Elbe and Ems.

> They weave nets of rushes and sedges to catch fish; and form mud with their hands, which, when dried in the wind rather than in the sun, is burned to cook their food, and warm their bodies chilled by the cold north wind.

There is archaeological evidence for peat-cutting in Denmark and both the English Fens and Somerset Levels in the pre-Roman period. Peat-cutting tools have been identified, and even cut turves over two thousand years old have been found. A Classical description of the Celtic Batavi tribe suggested that they were so wretched that '... *their drink is the drink of swine, and they burn their very earth for warmth*', meaning that they drank beer and burned turf. Early writers referred to peat as '*combustible earth*', so Cardinal Piccolomini, in 1458, described how the people of Friesland made '... fires of combustible earth, since they lacked fire-wood' (Rotherham, 2009).

Peat fuel use over the millennia has almost ceased in most areas. Many people associate peat-cutting with Ireland or the Western Isles of Scotland, or late-twentieth-century industrial extraction. Yet historically, peat and turf were the common fuels for most people across the whole country. Indeed the mark of the peat-cutter is written deep in the landscapes of Britain. In regions like Somerset, Cornwall, and North Wales, older people still recall the distinctive peat reek in their cottages. Associated with traditional community peat-cutting and use were distinctive tools and implements, and long-standing cultural attachments and folklore. In Scotland and the Isles, and in Ireland itself, peat-cutting is still important in culture, folk memory, and identity. However, across much of Wales and most of England, the memories are long past. I captured some of this in a book I wrote on the history of 'Peat and Peat Cutting' (Rotherham, 2009).

The cultural history of peatlands extends from the Iron Age to the twentieth century, which gave us the Norfolk Broads and other distinctive landscapes, and was a part of most people's lives. Towns and cities like York, Norwich, Kendal, Carlisle, and even Liverpool depended on peat fuel. The former mosslands are written into today's landscapes as place-names, even when all other evidence has gone. In recent times peat was used as fuel in the Fens, in the Somerset Levels, in the Lake District and North Lancashire, on the North Yorkshire Moors, in Devon and Cornwall, and in North Wales. In almost all these regions there is no active cutting today. There remain maybe three peat-cutters in North Yorkshire, and in West Yorkshire the Graveship of Holme is one of England's last organised community peat mosses, but there is a smaller site still being worked just north of here at Oxenhope.

How peat was cut, processed, and used varied through time and between regions, and this is a rich cultural tapestry that is in danger of being lost from memory. Local rural history museums preserve some implements and even buildings complete with traditional hearths but often run on a peppercorn income, sometimes when these close entire collections are dissipated. A deep sense of history and mystery concerns many of the traditions of common rights and their attachments to particular buildings or hearths – sometimes true and at other times wishful thinking. Famous cottage fires, or particularly those in old inns, burnt continuously for decades or even centuries and local people placed great importance on these.

Then in Britain, Europe, and the USA, from the mid-1800s commercial peat extraction and processing industries emerged – often the final nail in the coffin of many peatlands. Even so, the history and culture of the early industries mixed with older, local traditions and left an element of cultural heritage. However, industry and agriculture ultimately removed most peatlands as many were lost to drainage and farming '*improvement*' from the 1700s to the 1900s. Commercial peat-mining was often followed by conversion to intensive agriculture. Also, as peat as domestic fuel fell out of use, it was often displaced by firewood and then coal, gas, oil, or electricity – all more expensive but more convenient.

The Black Waters – peatlands as contested spaces

The connections between history and ecology are frequently overlooked. However, the human history of peatlands is central to understanding both their loss and in some cases their survival. Around the world, these sometimes vast landscapes were (and sometimes still are) rich resources for exploitation (from early subsistence to modern industrial usage). However, they were also places of sanctuary, non-conformism, and independence. Even today in Great Britain, since the removal of common rights and the imposition of extensive sheep-farming and grouse-shooting, many upland moors and bogs remain strongly contested landscapes. In the twenty-first-century peatlands, now often 'leisurely landscapes', the different recreational stakeholders vie for control and usage – from grouse-shooting to mountain biking, and from birdwatching to hill-running; moors, mires, and bogs are still contested.

A major reason for the loss of many peatlands was the displacement of long-established communities and their rights of common. In the uplands this involved ring-fencing the moors and bogs, depopulating them of indigenous people, and replacing them with paid shepherds and gamekeepers. In the lowlands, removal of the common people was generally followed by widespread drainage and often total destruction. The landscapes and their communities in these regions today are radically altered from those before the enclosure, to such an extent that in many cases the one-time existence of peatlands is totally forgotten. Similar parallels are seen around the world today, especially in the tropical peat forests. In these cases, the locally indigenous forest dwellers are aggressively (and often illegally) displaced as commercial interests pursue firstly the timber for logging and secondly the clearance of the peat forest and its reuse for plantation cropping.

There is a further twist in the human attitude to these peatscapes which is based around fear and loathing of them as vast, impenetrable 'black waters' full of dangers and disease. Bogs, fens, mires, and moors have been seen by outsiders as unpleasant and frightening places and their inhabitants often as strange, foreign people. Such fears and attitudes have shaped human approaches to peat landscapes past and present, and continue to do so even in modern times.

Peat and both bogs and fens are important in British culture; in the past valued but also feared with bogs as black waters, places of doom and gloom (Rotherham, 2013c). Neither land nor water, they are something in-between, and those not of bog or fen found them fearful places where a misplaced footstep could mean slow death by drowning. Almost everyone has seen the villain sink into the Great Grimpen Mire in Conan Doyle's *The Hound of the Baskervilles*, and many people still fear the northern moors as the place of the 'Moors Murders'. These places are productive landscapes that supported people for centuries, but fearsome and loathsome to strangers. Peat and bogs are imprinted on our minds and language with terms like '*bogged down*' and '*mired down*'. The idea of a '*damper*' or '*putting the dampers*' on something, or '*dampening down*' derived from placing a damp piece of turf on the peat fire the last thing at night to keep the fire smouldering until dawn. Seminal moments in history are linked to bogs, fens, and peat. It was in the bogs and fens of Sedgemoor and Athelney that the Saxon King Alfred held out against the Viking invaders, and the perhaps apocryphal cakes he burnt on the fire would have been oatcakes on a peat fire rather than the singed coffee cake of popular imagination. But this is an image emblazoned in the mind of every schoolchild taught English history, and at its core is a peat fire smouldering away. After 1066, Hereward the Wake, the last great Saxon leader, maintained his stronghold against Norman overlords in the Isle of Ely, a small area of dry land in a sea of fens, water, and peat bog. And the last major battle on English soil, the battle of Sedgemoor in Somerset was fought, won, and lost in the middle of a great fen.

In Ireland there are early references to peat in the few written sources to early Irish life. In particular there are extracts from the *Senchas Már*, the Old Irish law text. There were also significant fines for illegal peat-cutting, and cases of accidental drowning in '*the ditch of a turf cutting*' were excluded from liability. Most of the turf was cut as slices in a relatively dry and structured state. However, if the peat was too wet and problems of drainage too intractable to dry the working area, it was extracted as a sort of mud. In the bogs of Cavan, Leitrim, Down, and Sligo this was described by Boate in the 1700s: '*On that dry place where the mud is poured forth, sit certain women upon their knees, who mold the turf, using nothing else do it but their hands*'. This '*mud turf*' was still being made from peat taken from Ardee Bog in County Louth in the 1980s; taken down the River Dee to be sold as fuel in Ardee town (Rotherham, 2009).

Peat and peatlands

An account of peat and peatlands needs to address basic questions of 'what is peat?' and 'where is it found?' In describing peat it is necessary to consider the types as well as its physical and chemical characteristics, and these are along with how it is formed and how it behaves. In describing peatlands, that is, sites or landscapes where peat deposits are significant, there are issues of distribution, diversity, and classification to cover mires, bogs, marshes, heaths, moors, and fens. There is a rich literature and in recent decades, triggered by major works on restoration and a growing awareness of ecosystem service and economic value issues, there has been an explosion of papers and books on the subject. Furthermore, temperate peatlands are embedded in wider ecological landscapes and this is often not fully appreciated. So 'heathlands', 'moors', and 'commons' often share origins with peatlands such as bogs and fens, and indeed, maybe be physically embedded within them. This means that in both landscape historical and ecological literature there is a broad overlap and commonality with the writing of say Gimingham (1972), and Webb (1986, 1998), and for example, Rotherham (1995, 2011a, b).

Early accounts include the papers of Gorham (e.g. 1953, 1957) and then the landmark volume by Moore & Bellamy (1974) which in many ways was the start of the modern era of peatlands research. Their book was one of the first major accounts to attempt a comprehensive overview of temperate peatlands. Others followed and in 1995, for example, Wheeler & Shaw wrote a good overview of peatland classification and functioning. This volume also provided a first thorough account of approaches to restoration of damaged peatlands. Brooks & Stoneman (1997) produced a useful and succinct volume as a conservation management handbook for peatlands. Also, in 1997, Parkyn et al. edited a major multi-authored text on peatland conservation that brought together much of the current knowledge at that time. Bringing the outputs up-to-date, with many excellent papers over this period (such as Vasander et al., 2003), the huge undertaking by Joosten et al. (2017) was the most comprehensive account to date, though mostly restricted to Europe. Most of these works consider ecology, hydrology, chemistry, function, and restoration. However, almost none of these or indeed the extensive literature not mentioned here directly addresses the human cultural history and nature of these eco-cultural landscapes. Nor do they pay any attention to the archaeology of peatlands beyond the use of the palaeorecord to help reconstruct past landscapes and climates (see Buckland (1979), Buckland & Dinnin (1997), De Noort & Ellis (1997), Hodgkinson et al. (1995), and Middleton et al. (1995) and (2013) for examples of the depth and diversity of potential interest). The outstanding exceptions to this omission was the volume in 1981 by Godwin on the '*The archives of the peat bogs*' or that by Berry et al. (1996) '*Fenn's and Whixall Mosses*'; yet despite these publications there has been little or no transfer over to the mainstream ecological and restoration works.

When considering the broad remit of 'peatlands' the separation of temperate and associated mires and tropical peatlands is also an issue (see Page et al., 2002, 2009, 2011 etc., and Page & Rieley, 1998). Matters of tropical peatlands and the threats to them are currently in the forefront of carbon and climate-change issues and debates.

Peat is an organic material formed under more-or-less acidic, waterlogged conditions in bogs, and slightly alkaline conditions in fens of various types. Fenland peats may be far less acidic than bogs and may be slightly alkaline though this depends on the water source. With anaerobic status, dead plant material does not break down but instead accumulates to form a layer of partially decomposed organic remains. Over time, these may build from just a few centimetres depth to many metres. Around the world, peat has been growing for around 300 million years and prehistoric peat accumulations have been responsible for generating the fossil fuels we have today. Although peat formation varies with environmental conditions, we can make some general statements. It takes around one year to form one millimetre of peat, ten years for one centimetre, a hundred years for ten centimetres, and a thousand years for a full one-metre depth. Peat is made up of about 90% water and sphagnum moss, its main constituent at least for most temperate bog peats, can hold twenty times its own weight in water. In effect, peat bogs form enormous water-holding sponges. Sphagnum peat is rather acidic though this may be affected by the water entering the mire system, and the purest bogs are the so-called 'raised mires' which grow to form a raised dome. At this point the only water entering the system is from pure rainwater. These ecosystems are inherently low in available nutrients and the waters are acidic. Air pollution over the last two centuries has seriously compromised the conditions with both 'acid rain' (causing a drop in pH), and nitrates and other chemicals in the atmosphere leading to nutrient enrichment or 'eutrophication'. Both these impacts seriously disrupt the functioning of delicate ecological systems and cause imbalances in the peatland vegetation and fauna.

What is peat and how is it formed?

Peat is undecomposed vegetation formed under conditions of waterlogging or high rainfall. The moisture prevents air from getting into dead plant material and over time part-decomposed '*peat*' develops. The type of peat depends on the environment in which it formed and the plants from which it is made. Vegetation changes over time through natural progressions called '*ecological successions*' and so the type of peat formed in any one location changes. So a cut peat at one location, then a '*peat profile*', varies with depth; the lower profile is more fully decomposed and densely compressed. Peat formed in a fen or carr (wet woodland) environment is from reeds and sedges, relatively rich in mineral bases such as calcium, and is slightly alkaline; that from *Sphagnum* ('*bog mosses*') low in mineral bases is acidic. Lowland fen peats are the former and upland blanket bogs or mires, the latter.

The anaerobic often acidic conditions help preserve anything organic falling onto or into a peat bog. This explains 'bog bodies' in peat bogs and the less widely known, but scientifically interesting deposits of plant pollen and remains of insects from centuries or millennia past. These deposits are one of the main tools we have for 'reconstructing' images of past landscapes and ecologies to help understand long-term environmental changes. When peat bogs are destroyed this unique information, essentially an archive of past climates and ancient biodiversity, is lost.

Peat is influenced by the wider landscape in which it formed especially the rocks, groundwater, and surface waters. Mineral-rich waters from limestone affect peat composition, and coastal zones have a salty 'saline' influence. In some situations, now rare, peat-forming successions reach their ultimate pinnacle with great '*raised mires*'. Formed after a sequence of peat development on expansive, flat landscapes, undecomposed peat builds to such a point that it rises above the surrounding water table, so the only water input to the upper zones is pure rainwater. These systems are incredibly low in nutrients and develop a uniquely interesting but very vulnerable ecology. To walk across such a site with the consistency of a huge jelly is an amazing experience; neither land nor water the whole wobbly mass quivers and the trees shake. Most sites have gone, and those remaining are reduced to a fraction of their former extent (Figure 1.1).

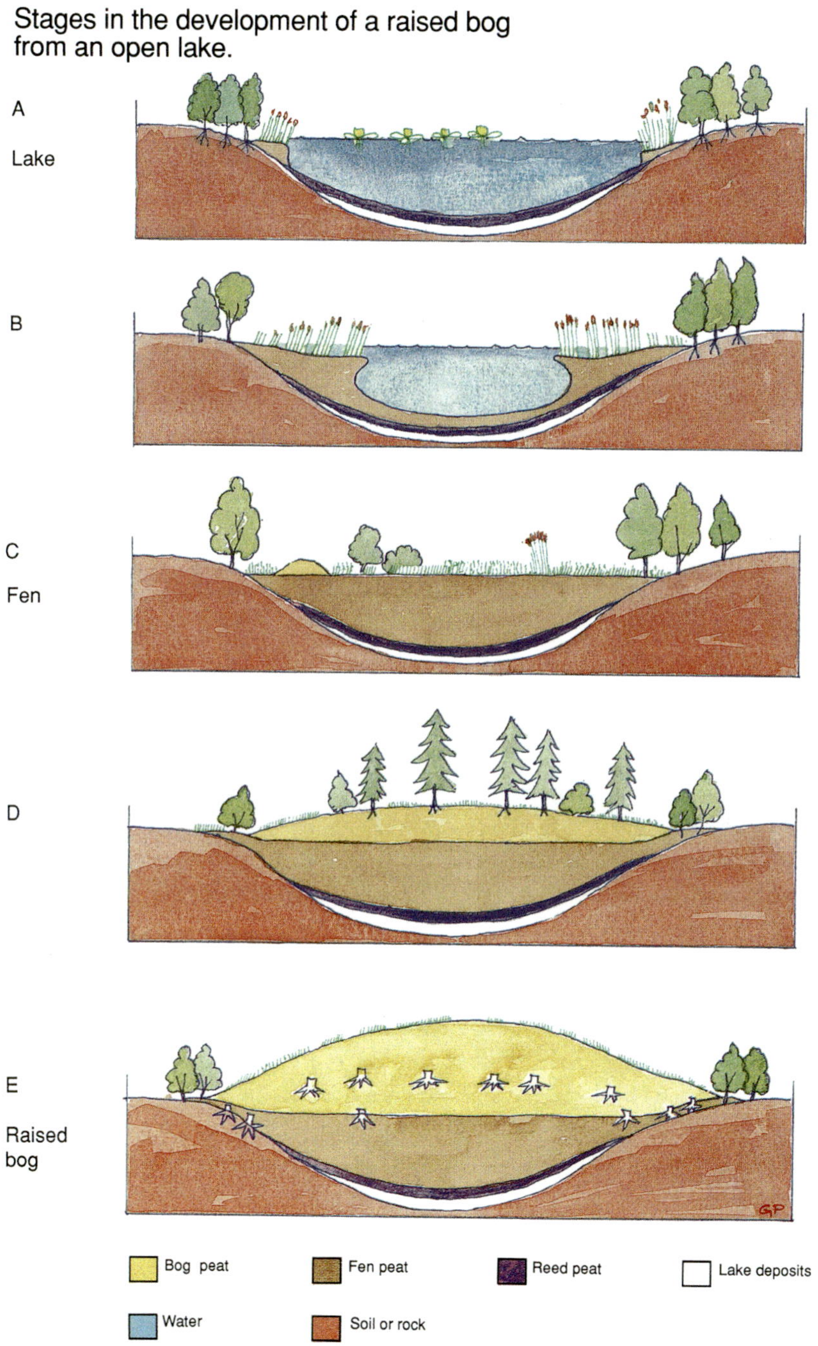

FIGURE 1.1 Development of a raised bog over time reproduced with permission from John Cross.

What are the peatlands?

By medieval times '*peatlands*' and '*commons*' of various types were noted on surveyors' reports as '*wastes*'. This doesn't mean they were unproductive; in fact far from it. These were wild, remote, but highly productive environments; sources such as the *Domesday Book* (1086) show lowland fens in particular, created considerable wealth. These were areas beyond farming cultivation with available technologies, and they were difficult to control politically. They were productive for local people but not for the Crown or big landowners. Over centuries, across much of Britain's rural environment, peatlands sustainably provided, fuel, food, and building materials for human populations, and grazing for domestic stock. The term 'waste' reflects perceptions of wealthier landowning classes that these areas were not productive for them directly, and by the nature of their terrain, were dangerous places. They harboured unrest and disease, and for those living outside the great fens, bogs, moors, or heaths, were strange and fearful places. Extensive lowland and coastal wetlands merged with riverine marshes and swamps, with rolling moor and heath from the lowlands up to higher grounds. All these had large areas of peat. For the ordinary people, the commoners and peasants, and for some landowners and later industrialists, this was an important resource. Given favourable conditions peat will grow on a site, but always slowly. So exploitation may mean the gradual eating away and reduction of the core resource. To gain access to very wet areas requires either drainage or dredging (as was employed extensively by Dutch peat '*miners*' in medieval times). Drainage results in drying and peat shrinkage and loss and facilitates agricultural encroachment, but may also result in landscapes flooding.

By the 1700s and 1800s, the demands for new and 'improved' agricultural land meant peatlands under direct assault, and in 200 to 300 years almost all those in lowland areas were totally obliterated (e.g. Rotherham, 2010, 2013b). The ones remaining are fragile and vulnerable environments, ranging from productive, lowland fens to low-nutrient, unproductive peat bogs, moorlands, and heaths. Subsistence use in remote areas of the Scottish Highlands and Islands was often not sustainable as local demand outstripped supply or the ability of the bog to regenerate itself; human suffering and depopulation resulted as they did in Western Ireland and the Irish Atlantic islands. Similar trends occurred in Cornwall and Dorset in England. In the depths of the cold medieval '*Little Ice Age*' these areas suffered abject poverty and peat was often their lifeline.

Where is peat found?

In the British landscape before widespread enclosures and '*improvements*', peatlands occurred almost everywhere. By early 1900, this had changed dramatically leaving a tattered rump of this once expansive resource. The most extensive British peatlands are now found in upland areas of moorland and blanket mire, from Dartmoor, Exmoor, and Bodmin in the south-west, to the Pennines along

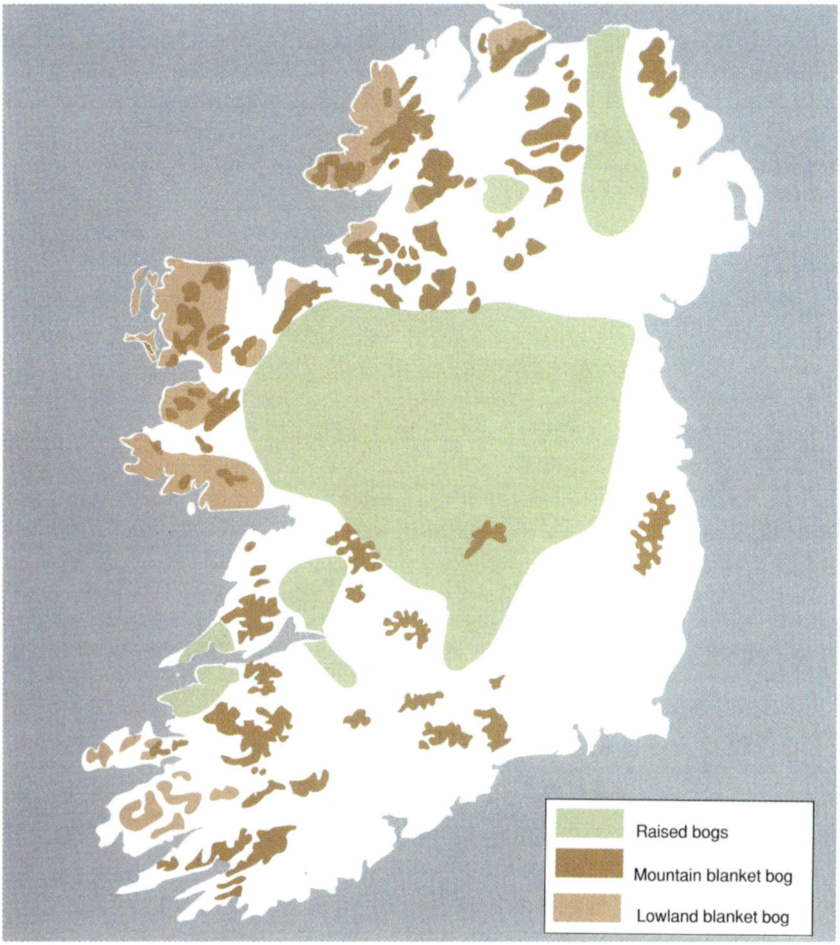

FIGURE 1.2 Types of bog in Ireland reproduced with permission from John Cross 1989.

England's spine, and then the hills and lower mountains of Scotland. Along the western seaboard, especially of the Scottish Highlands, in Ireland itself, and the islands around both these, are extensive blanket mires and raised bogs. Tracing the great river valleys, and in particular in their lower floodplains such as the Solway Firth and the Flanders Moss complex around the River Forth in central Scotland, are also the remains of once much greater peat bogs. In lowland and more southerly areas such as the Cambridgeshire Fens, the Cheshire and Shropshire Meres, and the Somerset Levels, there are relatively extensive peat bogs and fens. On drier heaths and moors are shallow peats and organic soils. Detailed historical research has shown former peat bogs and fens in areas that today would be almost unthinkable (Figure 1.2).

Peat or turf?

Confusingly there is no satisfactory definition of these two terms concerning the fuels and materials to which they refer. The site of peat extraction can be a turbary, a turf pit, a peat moor, a fuel allotment, or very often a peat moss. In many areas, the place-name of the local settlement or manor is attached to the local peat moss and this is the turbary for that area. For example, in Lancashire there are Leighton and Leighton Moss, and Silverdale and Silverdale Moss. The material itself is sometimes separated into '*peat*' for the deeper stuff, and '*turf*' for the shallower tops, but this is not always so. Sometimes '*turf*' is also used for the deeper peats. There are also localised and technical names with often subtle meanings in terms of brown fibrous peat from the higher levels and deeper, hard black peats from lower down. Turf is always the name used for thin heather and grass sward cut from drier sites and burnt, and the same implement, a turfing spade, was used for horticultural grass turf as well. These thin turves were generally considered poor fuel, but they did burn hot with high silica content from sandy soils, and in some areas were quite prized. Overall though, the vocabulary can be quite confusing.

The peat-forming plants

Most familiar peatlands such as in northern temperate regions have been developed by the long-term accumulation of sphagnum mosses, but other species (e.g. cotton-grasses, purple moor-grass) may also be important. Also, in fen vegetation in lowland regions the peat is from *Phragmites*, *Carex*, and associated vegetation. In southern temperate regions such as New Zealand or Tasmania for example, peat may be from sphagnum or mostly from other bog-forming species in waterlogged sites. In the Falklands, for example, peat may be from sphagnum mosses or a mix of grass and ericaceous species and simply develops where the wet, cool conditions slow the rates of organic decomposition.

There is a diversity of peat-forming species in the tropics for example, and in the southern hemisphere (as noted above). Tropical peat-formers vary from mangrove species in coastal zones to big trees in the swamp forests. These different peat-formers are introduced and described in the relevant chapters and their contributions to ecosystem services are considered.

Carbon storage

Contemporary peatlands contain around 500 gigatons of carbon which is about twice of that held in the global forests. Another often overlooked importance of peatlands is concerning water, with around 10% of the world's freshwater in peat sites. Furthermore, damage to peatlands has been responsible for increases in both flooding and drought. Unlike carbon storage in trees, woods, and forests for example, that in peat remains trapped in the soil system and safely locked away from the atmosphere. Generally, trees and other woody vegetation die and then decompose and the carbon is released again as carbon dioxide, though in peat forests, remains of the trees largely go on to form peat and the carbon moves into long-term storage. With peatlands, the carbon is held securely unless the site is drained and the peat breaks down. Indeed, one of the key problems for temperate peatlands in the twenty-first century is the history of drainage and associated over-intensive land use. Soil moisture content is a key part of the delicate balance of the peat bog.

Human impacts

Ever since humans began to farm the landscape and till the soil, they have sought to drain wetlands and to convert them to 'productive' uses. However, until the twin pincers of the 'agricultural revolution' and the 'industrial revolution', people generally lacked the technology and the energy or power to transform these landscapes (or more correctly, waterscapes) on a large-scale or permanent basis other than through incremental incursions over long periods. In some cases, however, exploitation of peat fuel, for example, took place very early and on a large scale. Dramatic examples include the English Norfolk Broads which were formed from medieval peat-cuttings for fuel to serve the ecclesiastical city of Norwich (Lambert et al., 1961) or the Peak District's Dark Peak which according to Ardron (1999) had even more peat extracted than the Broads at the same time (Ardron et al., 1996). Nevertheless, with the advent of modern industry and agriculture, people had the technology and energy to drain bogs and fens and to keep them dry. Furthermore, as artificial fertilisers were developed, they had the means to turn these reclaimed lands into productive agriculture.

Even in the Middle Ages, localised intensive grazing of upland moors and similar areas caused vegetation loss, soil compaction, and erosion. The latter can cut deep gullies into vulnerable peatland areas and it is easy for drainage downslope to be triggered and the process of degradation accelerates. By the time of the early industrial period, increasing levels of air pollution in the English midlands, for example, were exacerbating the dieback of the upland bogs, and sites were eroding catastrophically. Erosion and acidification can lead to the removal of vegetation and exposure of bare peat which then rapidly oxidises or erodes in a downward spiral of deterioration (Rotherham et al., 2004; Rotherham, 2014a).

Peatlands occur in diverse forms and support a rich and biodiverse ecology. In temperate zones, peatlands consist mostly of a range of bogs in acidic conditions and where water is 'shed' from the site, and fens, which are more base-rich and are sites into which water flows. The wildlife and vegetation vary accordingly from reed-bed dwelling bearded tits (*Panurus biarmicus*) in fens to insect-eating sundews (*Drosera* sp) in bogs. Along with these, though now dramatically reduced, are various wet woodlands or 'carrs' of willow (*Salix* sp) and alder (*Alnus glutinosa*). These were probably far more extensive and more widely distributed before the modern period of 'land improvement' from the 1700s onwards.

Cultural landscapes

The importance of peatlands as 'cultural landscapes' with a rich legacy of associated heritage and archaeology is often overlooked. Peatlands such as bogs and fens have supported communities and been exploited by them for thousands of years. Fens, in particular, are rich and productive landscapes as long as you have the knowledge and skill to use them. Both these habitats provided abundant fuel which was hugely important to early communities. Furthermore, they offered opportunities for grazing livestock from cattle and sheep, to massive flocks of domestic geese. In fenlands there was abundant small 'wood' from the carrs, reeds and sedge from the wetlands, and large numbers of fish such as eels. Bogs were less productive but the drier sites could be grazed and materials like rushes and moor-hay (*Molinia*) could be harvested. These activities, especially hand-cut peat harvesting, have left a rich heritage in a deeply cultural landscape. Fenland peat-cutting tools found in lowland England have been dated as over 2,000 years old (Rotherham, 2009, 2013b).

Furthermore, peat sites preserved 'bog bodies' with around 2,000 or more probably sacrificial executions retrieved just from the European bogs alone. The bodies date from around 800 BC to about 400 AD and are believed to be ceremonial sacrifices to the local deities. Bogs and fens also preserved ancient trackways of brushwood and coppice poles, along with settlement sites, and ceremonial 'henges'. Ancient boats and intact giant trees are also preserved within the deep peats.

The human uses of peat and peatlands are diverse and varied. For example, a largely forgotten use of peat moss or sphagnum has been the once-extensive application as surgical bandages – for application to wounds or in World War One, as absorbent pads for the widespread and distressing dysentery which debilitated so many troops on both sides of the conflict. Peat as 'peat litter' was also vital for stabling of warhorses during conflicts and especially on the Western Front in World War One (Rotherham, 2009, 2013b).

Drainage and destruction

In countries like Britain, the active drainage of fens, bogs, and marshes began as early as the Roman period. Then, from Saxon and Viking times to the later medieval, until their dissolution by Henry XIII, the monastic foundations coordinated much land reclamation, drainage, and improvement. From the 1600s onwards, coordinated and increasingly effective drainage and conversion of 'waste' to productive landscapes took place. Knowledge and understanding grew with industrial development and scientific discovery, and the demand for improved farming expanded with population growth and urbanisation throughout the 1800s and 1900s. In the face of this continuing attrition, the bogs and fens shrank and the ones which remained also changed from their more pristine condition of earlier times.

But it was the two World Wars that helped change rural landscapes so dramatically. Taking Britain as an exemplar case study, World War One triggered a timber crisis and in 1919, immediately after the war ended, the Forestry Commission was established with huge budgets and the task of recovering damaged woods and forests and also with wresting from neglect, large tracts of 'wastes'. The latter was particularly the extensive peat bogs and moors. Following World War Two, the forestry movement continued but now armed with American or American-inspired technology such as deep-plough drainage and tracked tractors to access and rip the most remote and intransigent sites (Rotherham, 2013a, b).

The governments of the day also funded private landowners and farmers to extensively 'grip' upland moors and blanket bogs to trigger the slow death of large areas of bog by progressive desiccation. In the post-war period, government grants funded the gripping of over 1.5 million hectares of British upland moor and peat bog to produce 'improved' farming lands. Hill-farm 'headage' payments were followed by EU Common Agricultural Policies which further exacerbated the problems of intensification throughout the 1980s and 1990s. The results for moors and bogs were devastating. However, having drained these lands with public funds for nearly a century we are now paying more public money for farmers and others to block the drains and to re-wet the landscape. What goes around comes around ....

Along with the drainage, has been massive over-grazing of upland landscapes and particularly by sheep. Again, subsidised by the public purse, farmers and landowners have had too many stock on the hill and for too long a period of the year. The results have been devastating for many upland regions and their ecology – now described often as 'sheep-wrecked'.

The extent and types of peatlands

Peatlands occur far more widely than is generally realised, and until modern drainage and 'land improvement' they were even more widespread. Worldwide, peatlands are to be found on all the continents, though much more in some than others. There is a problem in estimating either abundance or even loss since the figures are not comprehensive and are sometimes unavailable. Major 'undiscovered' areas of tropical peats have been found in recent decades. There are also issues of different definitions and classifications of 'peat' and 'peatland' which may vary from country to country. These may include bogs, mires, fens, moors, heaths, forests, and some grasslands and tundra. This too affects any assessments of occurrence and abundance. However, despite these impediments, good progress has been made in recent years to overcome the problems. In broad terms there are six or so major types of peatland that are widely recognised (as described by Sue Page at: www2.le.ac.uk/departments/geography/research/projects/tropical-peatland/what-are-peatlands):

1. **Blanket mires:** Rain-fed, generally 1–3 m deep. Location: mainly in Ireland and the UK. The UK alone has around 13% of the total global blanket mire area. They generally develop in cool climates with small seasonal temperature fluctuations and over one metre of rainfall and over 160 rain days each year.

2. **Raised mires:** Rain-fed, potentially deep peatlands. Location: Northern Europe and North America, as well as in the former USSR and parts of the southern hemisphere.
3. **String mires:** Flat or concave peatlands with a string-like pattern of hummocks. Location: primarily northern Scandinavia, but have been found in the western parts of the former USSR and North America.
4. **Tundra mires:** Peatlands with a shallow peat layer, *c.*50 cm thick. They form in permafrost areas in Alaska, Canada, and the former USSR, covering *c.*110,000 to 160,000 km^2.
5. **Palsa:** A type of peatland typified by high mounds, each with a permanently frozen core, with wet depressions between the mounds; found in the former USSR, Canada, and parts of Scandinavia.
6. **Peat swamps**: Forested peatlands (both rain- and groundwater-fed) are commonly found in high rainfall regions. This type of peatland covers *c.*350,000 km^2 and is found in South East Asia, Central Africa, the Amazon basin, and the Everglades in Florida, USA.

These are typologies mostly defined by depth, shape, water-source, and vegetation. But there are many variants and sub-types within these broad groupings. Peatlands occur in 175 countries or maybe more, and cover exceeding four million square kilometres or 3% of the Earth's terrestrial or freshwater surface, and form 10% of the freshwater resource. The greatest peat deposits are located in northern Europe, North America, and South East Asia. Those across South East Asia cover around 27 million hectares and with 23 million hectares in Indonesia alone, which amounts to 12% of the land area of that country. Indonesia has the most extensive tropical peatland and mangrove forests on the planet. However, estimating peatlands within the coastal marshes and mangrove swamps remains rather problematic! (See, for example, Page et al., 2002, 2011, and Page & Rieley, 1998) (Figure 1.3).

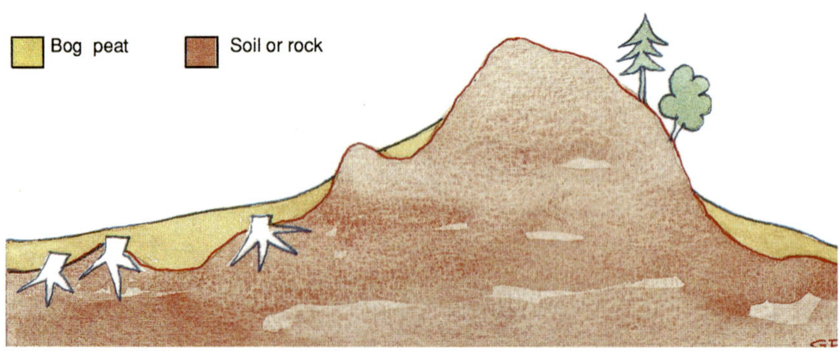

FIGURE 1.3 Section through a blanket bog in a landscape setting reproduced with permission from John Cross 1989.

The main occurrences of peatlands

The Americas

The most extensive areas of peatland are to be found in Canada (about 1,235,000 square kilometres), with other significant deposits in the United States particularly in Alaska. In both regions there are extensive zones of peatland habitats that have been removed by peat exploitation and by drainage since the 1800s. However, across the Americas around 82% of the total peat resource is held in the arctic, sub-arctic, and boreal regions in the extreme north and south of the continent. Every country in the Americas has some peatland and many of the islands too.

In the USA there are former wetlands along the eastern and western seaboards that have largely been removed. The drainage basin of the Mississippi in the central plains zone of the southern USA and the lower reaches were formerly dominated by extensive peat-swamp. East of this area is the extensive coastal peatlands of Georgia and of course, Florida has notable sites like the Everglades. The Prairies also have abundant but small 'pothole' peatlands.

Central America has little peat except for a few coastal swamps, and highland zone open bogs. In South America, the main deposits are tropical peat-forests in countries like Brazil and Peru and some upland bog areas in the headwaters of the great rivers. There are also extensive blanket mires in Patagonia. It is likely that both tropical and southern extents are under-recorded.

Britain and Europe

As a strongly Atlantic-influenced environment, with a predominantly cool and wet climate, Britain and Ireland have large areas of peatland, though much has gone and what remains is often seriously compromised. The UK, for example, has about 233,600 square kilometres of peat bog today but historically there was much more.

Across Europe, much of the peatland extent is along the Atlantic seaboard and then in Baltic and northern latitude zones. The most well-known European deposits are in the Scandinavian countries and across through Finland and the Baltic States, and Russia. Poland had and has extensive peat bogs that remain intact and Germany, France, and Holland also had extensive peatlands though many have been removed or badly degraded. There were extensive peat deposits in countries like Ukraine which have been extensively cut, and even in Turkey there are fen peats. Some of the European sites are open mire ranging from raised and blanket bog to reed-fen, and others have extensive tree cover.

Africa

There are small areas of peatland in many African countries and more extensive resources of tropical peat forests have been found in recent years. The most extensive deposits are tropical peat-swamp in places like the Congo basin, in high-altitude zones like Ethiopia, and southwards across Rwanda, and Burundi. Along the East African seaboard of the Indian Ocean, there are important resources including ancient sites such as the 45,000-year-old Mfabeni Mire in South Africa. Other areas of importance include the river deltas like the Okavango in Botswana and the Sudd in South Sudan.

Asia

Northern latitude parts of Russian Siberia have extensive peatlands and Russia has had a long-standing history of peat-mining which has damaged and destroyed large areas. These northern zones include extensive frozen peat bogs equivalent in area to France and Germany combined or thereabouts. However, this peatland resource is currently under threat because of enhanced global warming and it is actively thawing. Since this is the world's greatest storage system for trapped methane, which is a powerful greenhouse gas, its release could prove catastrophic. The amount stored is estimated to be around seventy billion tonnes of methane; around a quarter of the total methane stored on the global land surface. If this melting of the northern tundra peats continues, then the impacts on global weather systems may prove hugely damaging. The other deposits are found in China, and elsewhere in small amounts.

However, the most significant Asian sites are the tropical peat-forests such as in Indonesia and Malaysia, but again, these have been badly damaged and remaining areas are actively being drained, burned, and converted to palm oil plantations. Peatlands also occur on islands like the Philippines and around the coastal zones of the Indian subcontinent.

Australia, New Zealand, and the Antarctic

Peat is not so abundant in the southern hemisphere though limited resources are found in New Zealand, parts of Australia, and also on the Falklands and even Tasmania. Very small deposits are even found in the Antarctic. The vegetation and ecology in these southern peatlands is very different in terms of fauna and vegetation from the northern zones. Throughout the Oceania islands there are scattered smaller occurrences of peats. The peatlands vary from typical peat-swamp forest including mangroves to temperate bogs and fens.

Peatlands and climate

The wider issues of peat and peatlands have been discussed by many authors such as Maltby & Proctor (1996), Maltby (1997), and Bragg et al. (1992) (Figure 1.4). The links between peatland degradation and climate change issues were probably first stated with authority at the December 2007 United Nations summit on climate change held in Bali. The meeting stressed the increasingly urgent need to address the protection of the world's remaining peatlands and to restore

FIGURE 1.4 Amazonian tropical rainforest in a nineteenth-century print and illustrating the amazing diversity of peatland habitats.

those damaged but remaining. This probably did not take into account the extensive areas now being recognised as former peatlands though entirely removed from the modern-day landscape. It was suggested that remedial action could cut worldwide greenhouse gas emissions by around 10%. However, despite some good work, peatlands on all continents are still exploited unsustainably and many are destroyed entirely.

The consequences of long-term peatland degradation are now well-documented and there are excellent case studies that demonstrate human impacts. So, for example, on the moorlands of England's Peak District, there have been around 400 or so 'wildfires' reported since the late 1970s. Many of these are due to drainage of a combination of blanket bog and moorland, and the massive dieback of water-retaining sphagnum mosses due to air pollution from nearby big industrial cities. Some of the fires have burned for months and devastated areas of up to a thousand hectares or more. With climate change and increasingly frequent droughts, the risk of wildfires and subsequent rapid erosion and downwash of damaged peat is very high when heavy rainfall follows the droughts.

In good condition, a healthy and biologically active peat-soil acts as a significant carbon sink but this function is seriously compromised if peatlands are damaged or disrupted. Peatlands release carbon dioxide through aerobic respiratory activity as the water table is lowered and air allowed into the formerly waterlogged peat. However, the other major mechanism of loss that is less widely recognised is through erosion as described above, and that associated with footpath damage and deep gullying through into eroding deep peats. There is also major loss as dissolved organic carbon (DOC) is lost via the water running out from the bog along streams and gullies. Of this DOC movement, around 40% is returned to the atmosphere as carbon dioxide.

Work by Worrall at University of Durham (e.g. Worrall et al., 2004) has indicated that DOC levels in UK rivers began to rise significantly around fifty years ago, that is in the 1970s, and is increasing at about 6% per year. The causes of the losses, which seem to be occurring in nearly 80% of sites examined, are probably climate change and land-use intensification. During dry summers the increased temperatures and dry weather raise the rates of aerobic decay and associated carbon release into both the atmosphere and into the peat-water; much of the released carbon is then flushed out as dissolved organic material during rainy periods in the autumn and winter months. Furthermore, it seems that higher levels of atmospheric greenhouse gases actually cause bacterial decay of peat to accelerate and thus trigger an increased release of stored carbon. This in itself leads to enhanced climate change and a spiralling problem of associated change and release. With peat-soils holding around 30% of the total carbon stored in terrestrial ecosystems this is a major concern in terms of continuing environmental degradation.

Projects set in train over the last decade seek to repair some of the damage and to halt or at the very least slow further deterioration. Restoration to the point of active and sustainable carbon sequestration is perhaps a longer-term aspiration though flood water alleviation impacts can be relatively speedy. In the UK, there has been progress for example on issues such as the management of upland blanket mire burning associated with grouse-shooting.

Peatland restoration

The first volumes to address the restoration of damaged peatlands were probably Wheeler & Shaw (1995), Brooks & Stoneman (1997), and Parkyn et al. (1997). Recent publications such as Cris et al. (2011, 2014) have developed professional practice further. Active restoration of damaged peatlands may require fairly radical approaches and interventions. These include the raising of low pHs by liming, the introduction of heather (*Calluna vulgaris*) and associated plant species through the application of cut heather brash, the blocking by various means of the drains or 'grips' cut into moors and bogs and therefore holding rainwater onsite, and the introduction of sphagnum mosses by plug-planting or using cloned plant fragments in a bead of gel. Steep slopes of bare peat may need to be stabilised by use of textile covering such as jute.

Blockage of grips may be by the insertion of wooden or metal piling, using cut peat blocks dug out either by hand or by low-impact machine. In remote upland landscapes with difficult terrain this can be slow and laborious, and expensive. In lowland sites such as the Humberhead Levels National Nature Reserve in northern England, the restoration may necessitate the very gradual raising of the water levels over a wide and mostly flat area. The intention is to gradually re-create conditions suitable for growing the sphagnum bog-mosses and avoiding the making of an expansive lake where the wave-action generated by wind would simply erode any developing marginal vegetation.

Especially in upland areas, the aim is to restore peatlands such as bogs back to fully functioning biodiverse ecosystems, that take up and hold carbon dioxide, and which hold water like a sponge. This system can then help mitigate the effects of extreme weather events such as high rainfall and consequent downstream flash-flooding. In the UK, this approach can be part of wider measures to reduce flood impacts on vulnerable settlements such as the cities of York, Carlisle, Leeds, Sheffield etc. and of smaller settlements like Kendal and Hebden Bridge. All these have suffered catastrophic flooding in recent years. The carbon sequestration and storage is also long-term, whereas that held in trees may in part be returned to the atmosphere in a rather shorter cycle.

Historically, the ideas, techniques, and technologies for exploiting and destroying peatlands were largely developed in Western Europe and exported around the globe. Today, the new approaches to restoration are gradually been disseminated worldwide as peatlands are a truly global issue.

Bog bursts

Finally, one of the terrifying occurrences for those living near a major peat bog was the bog burst. This might be the result of the bog simply outgrowing itself to a point of instability or sometimes a result of climate. However, it was usually the consequence of human interference, frequently resulting from attempts to drain or to cut the bog for fuel. This activity could destabilise all or part of the bog with dramatic and devastating consequences. From the 1500s to the 1900s, there were numerous and spectacular bog bursts and slides, sweeping trees, houses, people, and livestock along with and them which would sometimes extend over several miles. Some of these (in for example 1526 and 1633), were documented for the Greater Manchester area by Hall et al., 1995).

2

GLOBAL PEAT RESOURCES

Peatlands worldwide

The extent of peatlands around the world is subject to issues of different definitions and interpretation of what exactly they are (Figure 2.1). However, there have been attempts to assess coverage and these have gradually become more refined and more reliable (e.g. Immirzi et al., 1992; Lappalainen, 1996a).

FIGURE 2.1 Industrial peat-cutting in Bavaria around about 1932.

Despite covering only about 3% of the world's surface, peatlands store 30% or more of the total global resource of soil-based carbon. This is a total land mass equivalent to the entirety of India and Pakistan combined and is mostly in the high latitudes and in some tropical areas. However, this distribution is uneven across the continents and there are big concentrations in the northern hemisphere across North America, northern Europe, into Siberia. Some tropical forest zones are also proving to be rich in peat resources. In some regions, the peatlands are dominant features in vast landscapes.

Estimating the peatland resource

An early report on peatlands and carbon was commissioned by Friends of the Earth from the Wetland Ecosystems Research Group at the University of Exeter, and published in June 1992 (Immirzi et al., 1992). On page 37 this report provides an overview and summary of the issues of producing estimates of peatland areas, including definitions, and gives a summary of previous estimations followed by a series of tables with data for continental and selected national audits. The review raises serious issues concerning the compatibility and definition of peatlands, but there are questionable assumptions too. For example, p. 38 notes the issue of estimates done to assess fuel potential but assumed that "Organic deposits with potential for fuel production require a minimum depth of 1 metre and more realistically 2 metres". From an historical point of view this can be misleading since thin surface deposits of organic matter, together with plant roots (grasses, heather etc.), were also extensively cut, harvested, and burned. Taking the strict definition for 'commercial' peat-stripping can, therefore, underestimate the real extent of areas affected. Furthermore, some estimates of peatlands are based largely on 'wetlands' and this has some logic. However, there has been considerable cutting of shallow organic soils on relatively dry sites and many now-desiccated landscapes were formerly wetter and can therefore be overlooked.

A pioneering paper was written in 1953 by Eville Gorham in the *Journal of Ecology*. In this he set out to investigate the history of the systematic study of peat and peatlands. At that time there was relatively little reference to work before the twentieth century. Furthermore, this seemed at odds with the long-standing and diverse language and nomenclature associated with various types of waterlogged habitats and soils. Interestingly, this was similar to when, in the late 1990s, I first researched the historic accounts of peat and its utilisation; this topic was almost entirely neglected by contemporary researchers on both sides of the Atlantic. The reader is directed to the 1953 paper for a fascinating insight into the neglected history of British peatlands. Eville's concluding remarks are closely in accord with my more recent perspective,

> … the nature, origin and development of peat lands. It appears that some of the basic facts and theories concerning such areas were established long before the development of ecology as a separate discipline by professional

> scientists. These early observations seem to have been very largely neglected during the modern period of research on bogs and fens.
>
> *(Gorham, 1953)*

Such lack of understanding of historical sources is central to issues of the former extent of temperate peatlands. This then relates to current debates on the implications of this, and of their removal and associated carbon dioxide release and climate-change impacts. Eville Gorham developed the ideas on peatland development in a 1957 paper. In North America, he considered peat and climate in a series of key papers such as Gorham (1991) on temperate peatlands and climate warming. Gorham and colleagues also gave detailed analysis of post-glacial carbon accumulation rates in North American peat bogs and predictions of future behaviour (Gorham et al., 2007, 2012). In brief, they conclude that northern temperate and tundra peatlands are major long-term sinks for sequestrated atmospheric carbon and this role is anticipated to continue. In many areas, the North America peatlands are less disrupted than those of Northern and Western Europe for example and in that context their ecosystem services are still reasonably intact. However, they also note the need to examine more fully the role and present underestimation of carbon in thin peats and this was considered in Rotherham (2020), for example.

Growing awareness of peat and peatlands

Smout (2000) observed that with the advent of the industrial and agricultural revolutions traditional values and uses of bog-lands were displaced by the concept of them as 'wastes'. On p. 94, he noted how "It was, of course, in the nature of improvers to describe the great watery places as wastes", and that

> the value of land at Hatfield Chase was raised from 6d to 10s an acre by the Dutch adventurers. But such figures obscure the worth of the place under water to those who lived there, for the marshes were, in their own terms, incredibly productive and attempts to drain them might be met as at Hatfield, by active and violent resistance by the inhabitants.

This theme is developed in later chapters. By the modern period of peat-related research beginning in the 1970s, authors such as Moore & Bellamy (1974), with their initial review became influential. Increasingly comprehensive reviews were provided by Kivinen (1980), and Kivinen & Pakarinen (1981). A review for The Friends of the Earth (Immirzi, & Maltby, with Clymo, 1992), noted difficulties and shortcomings in available datasets, and then relied on published sources to present a global figure of 386–409 million hectares of soils with peat over 30 cm deep and organic content over 50%. The figure is noted as being a lower band and conservative estimate. Such estimates were produced without modern remote sensing technology such as Landsat/GPS/GIS. The United Kingdom was considered to have between 1,400,000 and 1,600,000 hectares, and Ireland 1,180,000 hectares, though the figures excluded peats buried under alluvium.

In 1987, Burton & Hodgson estimated lowland mire in the UK to cover 100,000 hectares with a further 36,000 hectares in Northern Ireland (Hammond, 1979). Overall assessments using the Soil Survey of England and Wales gave a UK estimate of about 1.4 million hectares and similar to Robertson & Jowsey (1968) with 1.6 million hectares. Lindsay et al. (cited in the FoE report), also suggested just under 1.5 million hectares, with figures broken down by peatland types and national location (Table 2.1):

The Immirzi et al. (1992) report had a detailed appendix which considered issues in arriving at the figures at national and global levels. In the UK, the former area of peatlands was estimated to be 1,640,000 hectares with 894,000 hectares 'unaltered', and 747,000 hectares 'drained or otherwise altered', providing a figure of 54% unaltered. Nevertheless, these totals clearly underestimate the real levels of loss. In particular, there were serious difficulties in assessing damaged and disrupted peatlands. The 750,000 hectares given by Kivinen (1980) as "in agriculture (meadow land, pasture, and semi-natural grazings)" obviously assesses only part of the damaged resource. Likewise, Goodwillie (1984) estimated one million hectares that remained 'undrained' but again the values reached are problematic in terms of historical landscape transformation.

It has been estimated with very broad-brush figures that the east England fenlands alone had 2,000 to 3,000 square kilometres of wetland in the Northern Fen and 3,000 to 5,000 square kilometres in the Southern Fen (Rotherham, 2010, 2013b). Furthermore, these areas of peatland habitat were present up to the 1600s and mostly gone by the 1900s. Such landscapes probably included mixed wet heath, raised mire, wet forest/woodland, peat fen, silt fen, and a range of marshland types. Taking a lower band of say 5,000 square kilometres hectares total, that is about 500,000 hectares in just this region alone, would by itself exceed the 1992 report total for England and Wales by threefold. If we add to this the extensive lost wetlands along major river valleys (the Severn, for example), the Somerset Levels, the Shropshire, Cheshire, and Lancashire meres and mosses, the figures increase the estimates of the 1992 review. With extensive

TABLE 2.1 Based on Lindsay cited in Immirzi et al. (1992) approximate areas of former mires in the UK [with figures re-calculated and totals corrected]

Bog type	*Area in hectares*		
	England and Wales	*Scotland*	*Total*
Lowland raised mire	37,000	24,000	61,000
Blanket bog	225,000	1,030,000	1,255,000
Intermediate bog	4,900	8,200	13,100
Fenland soils	148,800	520	149,320
Total	415,700	1,062,720	1,478,420

upland blanket mires, Ardron and collaborators (1996, 1997, 1998, & 1999), demonstrated a huge contraction in the South Pennine peatlands over around a half a millennium. In this case, even today, the wider peatland extent is mostly unrecognised and the total removal of peat to the mineral soil means many of these former areas are overlooked by contemporary surveys.

Carbon and peatlands

The most thorough review of carbon and peatlands is by Lindsay (2010), which provides an exhaustive evaluation and succinct summary of research and issues. Essentially from this starting point, the importance of peatlands to carbon cycling and storage is well understood and this is complemented by studies on soil carbon and carbon cycling more widely (e.g. Heinemeyer et al. in Kutsch et al., 2010). There are then questions about the former extent, nature, and depth of peatlands and associated habitat types and the implications for carbon capacity.

Taking temperate peatlands as a specific example, there is considerable literature (e.g. Lindsay (1992, 1995, & 2010) and Anon (1994)). Recent publications include Bonn et al. (2016) and Joosten et al. (2017), and standard texts like Charman (2002) give comprehensive overviews of peatlands, their ecological landscapes and issues of degradation and restoration. These publications consider impacts of peat-cutting but not on the scale suggested by historical research (Rotherham, 2020). Bragg et al. (1992) noted human impacts on peatlands and restoration, and Cundhill & Whittington (1992) researching Fife in Scotland, commented on landscape impacts of peat-cutting for fuel noting how "The historically recorded large peat mosses (e.g. Mossmorran and Star Moss) are today only shadows of their former extent and all reveal evidence of peat removal". In such cases, peat-cutting occurred until all the peat was removed, or like the Norfolk Broads the workings became unusable because of flooding. Also, many such worked peat areas are now masked by more recent sediment deposition and can easily be overlooked. As they stated, "At the present time there are few obvious bogs in Fife and one of the major reasons for this is the exploitation of peat as a fuel resource in historical times". Today's absence of peatland sites is because of historic peat-cutting for fuel (Figure 2.2).

Meade et al. (1997) included work on the extent of historical upland peat removal for industrial and commercial interests, and domestic fuel. Evans & Warburton (2007) provided a comprehensive overview of the upland peatland topography including erosion impacts but omits any reference to widespread peat-cutting and associated drainage. A conclusion from this brief overview is that the recognition of the one-time extent of temperate peatlands and of how much has been lost thereafter is very limited. The real extent of these landscapes was considerably more and therefore the damage to the ecosystem services and the carbon release, for example, are far greater than presently perceived. This sits alongside the realisation that tropical peats for example are far more extensive than previously understood.

FIGURE 2.2 Massive removal of peat by industrial exploitation in Belgium in the 1930s.

The situation for temperate peatlands and particularly issues such as carbon sequestration are developed further in Rotherham (2020). Taking an historical view this argues that the historical reduction of the extent of temperate peatlands has been far greater than currently recognised and this has significant implications for the release of carbon dioxide into the atmosphere at a relatively early date. This observation has relevance to debates on the triggering of global climate change. Furthermore, the extent of the contraction and shrinkage of temperate peatlands impacts on any potential roles to mitigate future environmental changes. However, as discussed with Andreas Heinemeyer (pers. comm.) the observation also carries a positive message which is that there may be far greater potential to re-establish peatlands into the landscape to ameliorate atmospheric carbon dioxide through future sequestration.

Carbon and the global peat resource

With peat, peatlands and other wetlands are significant in the carbon cycle and global carbon sequestration. Lappalainen (1996a) brought together key information on peat resources and whilst individual country-based contributions may vary this does provide a helpful historical overview. The estimates provide for pre-disturbance mass of peat-related carbon fixation as between 330 and 530 Gt C in peat-soils compared with 1,600–1,800 Gt C stored in far more extensive non-peat soils. Carbon stored in peatlands is around 20% or more of the total soil-stored carbon and about three times that of tropical rainforest biomass. (Recent discovery of extensive tropical peat-forests will change these figures significantly to increase the peat-related component). Peatlands not only accumulate carbon but they do this over long periods of time and storage is long-term unless the sites are disrupted. These are 'unbalanced' systems where carbon in dead organic matter accumulates more rapidly than it decomposes so layers of wet organic material build up to form peat. Mire ecosystems sequestrate atmospheric carbon and store it long-term. They can absorb about 100 million tonnes of carbon per year. This is a major contribution to carbon balance and thus to climate-change mitigation.

However, with so many peatlands (mires, marshes, and fens) drained, desiccated, and cultivated for agriculture and forestry, these landscapes have become net contributors to global atmospheric carbon rather than functioning as sinks. This calamitous change has major implications for climate change processes. Indeed, it is suggested that around 600 million tonnes of peat-stored carbon are now released back to the atmosphere annually because of human disruption of peatlands.

The role of peatlands concerning climate change is complex with issues such as carbon dioxide flux, carbon sequestration, and, for example, methane release from older bog systems. Maltby & Proctor (1996) considered the human utilisation impacts over the last two centuries noting about 300,000 square kilometres impacted upon, with human attrition growing at around 5,500 square kilometres annually. They acknowledged lesser contributions to carbon release from forestry and pasture on peatlands but highlighted the major effects of arable farming (41 t $CO_2ha^{-1}yr^{-1}$ from converted temperate peatlands; 154 t $CO_2ha^{-1}yr^{-1}$ from converted tropical peatlands).

For some reason which remains a mystery, carbon release from peat-soils is presently not factored into national carbon-accounting calculations relating to climate change and international protocols. Lappalainen further asserted that the contribution of peatlands in terms of carbon flux is significantly underplayed in global carbon and climate-change models. Maltby & Proctor noted the important effects of 'traditional' utilisation and management of the landscape in shaping many regions such as the expansive wetlands of the English Norfolk Broads or the Dutch Plassen – both flooded, abandoned peat-cuttings. Lappalainen gives figures for total global peatlands covering 3,985,000 km^2 and other wetlands 2,428,000 km^2.

The Baltic countries have rich peatland resources but also long periods of exploitation and human impacts. Bambalov (1996) gives Belarus figures of 9,631 km^2 drained for agriculture, a peat reserve of 1,575 Mt, 1,834 km^2 cut-over peatland, and 1,090 km^2 now actively involved in industrial 'peat-winning'. Other lands are nature reserves with some described as reserves of 'valuable' peat types, and 'other areas' amounting to 7,975 km^2 with 1,479 Mt of peat reserves. About 42% of Belarus peatlands were converted to agriculture, forestry, peat extraction, and urban or industrial development. In Estonia Orru (1996) noted around 2,580 km^2 drained and used as pasture, grassland, and forestry. Nationally there were 5,300 km^2 of peatland with 500 km^2 protected. For Latvia, Snore (1996) suggested 15% of the peatlands were now agricultural, 8% forestry, 1% 'ponds' (presumably carp fisheries), 4% commercial peat extraction, 2% cutaway, and 70% currently 'unused'.

FIGURE 2.3 Huge Russian peat harvesters operating across extensively cut peat bogs in 1960.

Another further example is Norway where Johansen (1996) gives 1,900 km^2 converted to agriculture, 4,100 km^2 drained for forestry, 270 km^2 used for peat fuel, 25 km^2 for other peat-industry uses, and about 1,000 km^2 protected areas and reserves (Figure 2.3).

In Great Britain, Burton (1996) gave overview figures indicating total peatland area of 17,549 km^2 plus 31,070 km^2 shallow peats and mineral soils with peaty horizons. This review gives more ambitious figures than the 1992 FoE report which reflects a broader and more inclusive assessment. Resonating with figures noted earlier (Rotherham, 2010, 2013b) for the east England fens, Burton suggests that the Southern Fen basin (i.e. the East Anglian Fens) was formerly the largest British lowland mire with around 1,480 km^2 in about 1630 AD. This date was before the start of the great drainage and 'improvement' to agriculture. By the 1980s, the functioning peat landscape was reduced to just a few, scattered, tiny areas with peat-based soils diminished to less than 240 km^2. Furthermore, all the remaining areas have been subject to massive peat wastage from desiccation, intensive agriculture, etc. Today fewer than 106 square kilometres have a peat depth of more than one metre.

With extensive areas now at or below sea level, raised mire peats are restricted to about 11 km^2 centred on Holme Fen which is itself drained. Before recent 'restoration' works, sites of relatively intact fen at Wicken, Woodwalton, and Chippingham totalled around 600 ha. Recent major projects have sought to restore parts of the 'Great Fen Project' around Holme Fen, and the 'Wicken Fen Project' area. These are ambitious by nature conservation standards but relatively tiny when compared to the historic losses. Beyond former raised mire zones are soils derived from peat fen and silt fen which amount to about 1,240 km^2

of variable peat depths which are mostly superficial and rapidly dissipating. The situation in other former fens and lowland mires like the Somerset Levels, the Shropshire, Cheshire, and Lancashire meres and mosses, Thorne and Hatfield Moors in South Yorkshire, Flanders Moss in Scotland, and elsewhere mirrors this general pattern. Nevertheless, the east England Southern Fens provide a stark example of the scale of loss and the nature of what might survive following 400 years of intensive 'reclamation' and 'improvement'.

Interestingly, when we consider the various national and regional accounts of peatlands and their utilisation and history, there is little on domestic peat extraction and its long-term impacts. As noted earlier, this again suggests that many figures are significant underestimates of former extent. However, as a corollary to this observation in terms of future restoration and carbon-sequestration projects, the areas of land for potential recovery may be far greater than currently indicated (Figure 2.4).

FIGURE 2.4 Peat cut as fuel in Ireland in 1947 in response to winter cold and European coal shortage.

Biodiversity and peatlands

In some regions like northern Europe, North America, and Asia, peatlands are dominant environmental components hugely important to wildlife, biodiversity, and ecosystem functions. The mires range from expansive, often treeless blanket bogs, to deep-frozen permafrost areas, to forested bogs of, for example, the Siberian taiga, and lowland raised mires. Much tropical forest which remains relatively intact is on tropical wood-peat.

Globally, mires contribute significantly to biodiversity with much associated with the specialist conditions; waterlogging, low levels of available nutrients, and often acidic conditions, combine to support remarkably specialised floras and faunas. European peat bogs, for example, have insectivorous plants like sundew (*Drosera*), butterworts (*Pinguicula*), and bladderworts (*Utricularia*). Other distinctive bog species include bog myrtle (*Myrica gale*) bog rosemary or Andromeda (*Andromeda polifolia*), bog asphodel (*Narthecium ossifragum*), and cotton-grass (*Eriophorum* spp). In tropical areas, insect-eating pitcher plants (*Nepenthes* spp) are found in forest peatlands, and in North America there are *Sarracennia* species with funnel-shaped leaves holding enzyme-rich water that digests insects. Many other species such as breeding birds occur predominantly on peatlands but also occur on other habitats or else seasonally. These include birds such as dunlin (*Calidris alpina*), redshank (*Tringa totanus*), golden plover (*Pluvialis apricaria*), curlew (*Numenius arquata*), and common snipe (*Gallinago gallinago*), that breed on moors and bogs but in winter on and around lowland or coastal areas. Birds like skylark (*Alauda arvensis*) which in Britain have generally declined in the wider countryside, still do well on raised mires and even on damaged peatlands. These habitats, especially moors and blanket bogs, are also excellent places for birds of prey and owls, such as merlin (*Falco columbarius*), hen harrier (*Circus cyaneus*), and short-eared owl (*Asio flammeus*). The areas of peatland which remain, hold on to species that formerly inhabited vast areas of 'unimproved' landscape. Some of these unimproved and unenclosed landscapes were wet mires, but others were essentially dry heath and common (Rotherham, 2011a, b, and 2013b, and see Webb, 1986, 1998, or Gimingham, 1972, for example).

In many countries, mires which survive may form almost the only available wildlife habitats away from degraded agricultural, urban, and industrial areas. Tropical peat-forests are famously rich in unique, and in many cases as yet undiscovered, biodiversity. In Indonesian and Amazonian peat-forests, many sites are being cleared with unknown species being lost without ever being 'discovered' or described by science.

Peat bogs can mix ecological and cultural facets to provide resources and reserves for both. They hold remarkably special plants and animals, provide refuges for species pushed out of adjacent habitats, places for typical wetland wildlife, and may be a significant component of regional biodiversity. Through this interest, peatlands can be significant contributors to the regional tourism economy. In Northern Ireland, for example, Peatlands Park (in County Armagh) has become the region's second most popular tourist attraction with many thousands of visitors each year. Such conservation and education initiatives have now been established at many sites throughout the world, though some reserves are too vulnerable and sensitive to cope with visitor pressure.

Bog sites may be especially problematic in terms of occurrences like trampling damage, in which case if access is allowed, then provision of constructed board-walks may be necessary. Visual disturbance can also be an issue for breeding and roosting birds in open bog landscapes, but again, impacts may be mitigated by strategic placement of screens and hides. Tropical peat-forests are also home to iconic and threatened species, such as the orangutan (*Pongo* spp), and appropriate visitor facilities can boost tourism income whilst also helping promote vital conservation.

In some cases, major, expansive areas of peatland can provide an important refuge for nature and at the same time, be places where people can experience a degree of 'wilderness' that is absent from most places in the modern world. Vast, open bog-lands of Siberia, of northern Scandinavia and around the Baltic, of Canada and Alaska, for example, are both home to indigenous peoples and places that others wish to visit. In Scotland, the Flow Country of Caithness and Sutherland, and the great plateau of the Cairngorms, provide similar opportunities. Rare breeding birds in such areas include red-necked phalarope (*Phalaropus lobatus*), red-throated diver (*Gavia stellata*), black-throated diver (*Gavia arctica*), and wood sandpiper (*Tringa glareola*). The great, tropical peat-forests, though under great threat today, offer a remarkable taste of a wilder nature.

History and heritage in the peat archive

Although we value peatlands as wildlife areas and as 'wilderness' they actually show evidence of human impacts and occupation over millennia. Furthermore, the same conditions of waterlogging and anoxic sediments that trigger peat formation also preserve human artefacts and even bodies. The anaerobic conditions prevent the growth of bacteria and fungi that would otherwise decompose organic materials. These form the archives of the peat bogs along with the preservation of other remarkable materials throughout history and prehistory such as pollen grains, seeds, fruits, nuts, and leaves from plants, the remains of the hard casings of insects, tree trunks from lost forests (generically called 'bog oaks' but often oak, pine, or even yew), and layers of sediments from periods of coastal flooding and saline inundations, or from perhaps muds and silts from great rivers. Preserved layers of sediments, peats, and other artefacts can be 'read' like a book to tell scientists about past times – their conditions, climates, and biodiversity. Tephra from volcanic eruptions, carbon-dating of materials and sediments, and tree-ring analysis from wood and timber, can all help provide reliable dates for features and activities. Further information may be gleaned from materials like snail shells or from micro-organisms like diatoms or foraminifers (unicellular algae and fauna) whose silica-rich cell walls survive in the peat. Essentially, pollen and other small particles in the atmosphere deposit onto the flat surfaces of wetlands or mires, for example, and are then incorporated into the sediment layers (e.g. peats, muds, or silts). Under waterlogged conditions these materials are preserved *in situ* and in the time sequence in which they were laid down. If the sediment accumulates over centuries or millennia, then the different layers can be dated and analysed.

Other examples of features preserved in this way include prehistoric trackways across great bogs and fens, preserved 'bog bodies' that are now believed to be from ceremonial sacrifices perhaps 1,500–2,000 years ago, dugout canoes, and wooden 'henges' in wetland sites. Layers of charcoal in the peat profiles may tell of human occupation and burning of the landscape and may correspond with tree clearance and groupings of bog oaks. The palaeo-environmental record preserved in the peat bog, both natural and anthropogenic, can help provide new interpretations of landscape change and human activities. In some cases, such as Flag Fen in Cambridgeshire, sites were clearly of religious significance with high-status offerings to local deities. Human sacrifices such as at Lindow Moss in Cheshire are rather more disturbing for modern cultural perceptions.

In part, the archaeological archive relates to people drawn to wetlands to settle and exploit the riches of available wildfowl, fish, peat fuel, and building materials. However, the remains also derive from trackways and routes across what were otherwise dangerous environments though which to travel; examples include the Sweet Track from around 3806 BC in the Somerset Levels. Well over a thousand such tracks have been discovered in peatlands across Ireland, Great Britain, The Netherlands, Germany, and Denmark. These and other associated features provide evidence of sophisticated building techniques and woodland management by the early Neolithic Period. Similarly, wooden log boats have been found in North America and northern and western Europe.

As early as the Neolithic Period, communities were moving through wetlands and along rivers and coastal zones in large dugout canoes. Some Iron Age settlements, such as lake villages in the Somerset Levels at Glastonbury and Meare, were located within great wetlands, but most people lived on the drier peripheral areas. All in all, the peatlands provide remarkable and unique insights into human activities over thousands of years and into environmental changes as well.

PART 2

Ecology, biodiversity, and hydrology

3

TEMPERATE PEATLANDS – THEIR ECOLOGY, WILDLIFE, AND FUNCTIONING

An introduction to the ecology and functioning of temperate mires

The extensive literature already noted, including major works such as Joosten et al. (2017), provides a good insight into the ecology and functioning of temperate peatlands. Peatlands and associated habitats are wonderfully rich places for wildlife, species that are rare and in these habitats sometimes abundant. The vegetation and wildlife vary considerably with the different types of peatland and with biogeographic factors such as latitude and altitude. Fens and bogs hold radically different ecological communities, and associated habitats include rich sites such as wet heaths, marshes, and carr woodlands. Upland peats are important for breeding birds, notably waders and birds of prey, whilst lowland fens hold bitterns (*Botaurus stellaris*), reedlings (*Panurus biarmicus*), marsh harriers (*Circus aeruginosus*), and much more. Both these examples of peat habitats have rich insect faunas with abundant and distinctive dragonflies and damselflies, Lepidoptera (such as the swallowtail (*Papilio machaon*) and large heath (*Coenonympha tullia*) butterflies), and both rare beetles and spiders. Superabundant crane-fly (tipulid) larvae on upland bogs and wet heaths provide essential food not only for birds like grouse (*Lagopus lagopus scotica*) and lapwing (*Vanellus vanellus*), but also for summering starlings (*Sturnus vulgaris*). Wintertime brings a whole different avifauna to peatland habitats, both fens and bogs.

Peatland vegetation (Figure 3.1) is influenced by both the type and condition of the site with considerable difference between uplands and lowlands, between bog and fen, and with factors such as grazing and mineral flushing. Levels of available nutrients have a considerable effect on peatland floras and so better quality blanket mires, the richer fens, and raised bogs there are often distinctive and rare species such as sundews, butterworts, bladderworts, bog rosemary, bog myrtle, and various orchids.

FIGURE 3.1 Sphagnum mosses in the Burbage Valley of the English Peak District.

This chapter provides an overview of the ecology of the various peatland types and the key roles of the peat-formers such as sphagnum mosses. The typical vegetation types of temperate peatlands are introduced and discussed and examples are given of the typical wildlife of temperate peatlands such as in Britain. This is along with an explanation of vegetation classification systems concerning peatlands and the recognised international classification of peatland types.

There are three broad types of peatlands in Britain and these are:

1. Blanket bog;
2. Raised bog; and
3. Fen.

Within these three broad types there are numerous variations and subtypes which relate to topography, geology, local hydrology, climate, and pollution. Human exploitation has dramatically changed the condition and extent of many sites. Despite this, today in Britain, around 70% of our drinking water originates from peat-dominated landscapes. Additionally, peatlands help provide water quality enhancement to benefit drinking water and other users, and they influence water supply by mitigating both excess (flood) and shortage (drought). Most important perhaps, is that intact peatlands help 'slow the flow' and so protect downstream, often lowland, towns, cities, and farmland from flooding.

Blanket bogs and lowland raised mires are both threatened habitat types at a global level. Europe has a significant proportion of the world resource and the UK, for example, has around 13% of the known world blanket bogs. Overall, however, the UK and Ireland rank in the top ten regions or countries in terms of peat resources. The UK, for example, has around two million hectares of blanket mire of which 60% are in Scotland. This amounts to a vast accumulation of peat which stores more than three billion tonnes of carbon, that is, the equivalent of all the forests of France, Germany, and the UK combined.

There are different approaches to describing the various types of peatland or mire ecosystems. Volumes such as Brooks & Stoneman (1997), Parkyn et al. (1997), and Wheeler & Shaw (1995), give useful overviews of these systems. Furthermore, the IUCN Commission of Inquiry (2011) offers a broad review of relevant policy issues too. Early classification schemes of temperate bogs based on work in Britain and Europe, split peatlands into continental, Baltic, Atlantic, and upland raised bogs. Baltic raised mires are considered to be classic examples when they have an outer 'lagg' zone, a sloping 'rand', and an inner 'cupola'. It was also observed that a lot of the Atlantic forms of raised mires tended to be flatter-topped with more of a plateau than a dome across the open mire surface. Such 'raised' mires are generally discrete areas with a clear perimeter boundary at the rand. Unconfined peat bogs occurring over an undefined area are known as 'blanket bogs' or 'blanket mires'; but these may include 'raised elements' within them.

Mires and bogs can be further subdivided by describing their broad shape in the landscape. These may be 'concentric' (effectively symmetrical), or eccentric (i.e. unsymmetrical). Most British raised mires are concentric though a few are eccentric, and of course, few if any pristine raised mire systems remain intact in Britain.

Bog vegetation develops under conditions of waterlogging, acidic status, and low levels of available nutrients. This leads to distinctive vegetation which varies with climate, topography, hydrology, and ecological succession. Distinctive plant species include the sundews and butterworts associated with the trapping and digestion of insects to supplement nutrient supply such as nitrogen. At a continental scale, such as across Eurasia there is a wide range of variants of temperate mire systems. Some, such as British raised mires are generally without tree cover, others like the eastern continental bogs such as in south Germany, are forested. Many North American bogs and swamps are also treed.

British bog systems vary north-south with a range of conditions from hyper-oceanic (Atlantic) to oceanic climates, and with altitude. In terms of the 'bog' elements of British mires, for example, there are four broad groupings and types (see Brooks & Stoneman, 1997):

1. Western blanket bog
2. Eastern or high-level blanket bog:
3. Lowland bog
4. Damaged bog

The last one of these (number 4) is the consequence of major human impacts on one of the others (1 to 3) and may vary from being slightly disrupted mire to being almost completely destroyed. The forms and conditions of sites regarded as 'damaged bogs' are discussed later.

Western blanket bog

This type of mire system occurs in very wet or hyper-oceanic environments which have more than two metres of rainfall per year and over 200 wet days. Sites are mostly below 500 metres altitude and frost-free. The vegetation has much purple moor grass (*Molinia caerulea*) and deer grass (*Trichophorum germanicum*), with *Sphagnum papillosum* and *S. capillifolium*, together with hare's tail cotton grass (*Eriophorum vaginatum*), and heathers and heaths such as *Calluna vulgaris* and *Erica tetralix*. In very wet zones there may be more sphagnum and common cotton grass (*Eriophorum angustifolium*), together with bog asphodel and *Sphagnum denticulatum*. There may also be bog myrtle, black bog-rush (*Schoenus nigricans*), and white-beaked sedge (*Rhynchospora alba*). Drier parts of sites may have fewer sphagnum mosses but have *Racomitrium lanuginosum* in abundance together with a diversity of *Cladonia* lichens.

Eastern or high-level blanket bog

With rainfall between about two metres and 1,200 mm per year, and from 160 to 200 wet days, the blanket bog vegetation which develops is typically dominated by heather and hare's tail cotton grass. In Britain, this is widespread across higher altitude and eastern areas such as the Scottish Eastern Highlands, the Southern Uplands, and the English Pennine moors. The high-altitude sites have plants like cranberry (*Vaccinium oxycoccos*), cowberry (*Vaccinium vitis-idaea*), cloudberry (*Rubus chamaemorus*), among others.

However, in some areas such as the South Pennines, the air pollution from cities and nearby industries has severely affected the vegetation and the peat. Combined with impacts of destructive wildfires, drainage, and intensive grazing with sheep, this meant that some areas have been transformed and are dominated by cotton grass and bare peat with the almost total loss of heather (*Calluna vulgaris*).

Lowland bog

In Western European zones with less oceanic influence, the bog vegetation is generally raised mire dominated by carpets of sphagnum mosses with *S. magellanicum, papillosum, capillifolium, tenellum, cuspidatum*, with *imbricatum, fuscum*, and *pulchrum*. Common flowering plants include heather, common cotton grass (*Eriophorum angustifolium*), hare's tail cotton grass, cross-leaved heath, cranberry, round-leaved sundew (*Drosera rotundifolia*), and bog rosemary. These communities are predominantly on raised bogs but may also occur in parts of blanket mires systems like saddles and in deeper hollows. *Sphagnum recurvum* and *S. cuspidatum* are typical of the wetter zones in these habitats.

Damaged bog

By the twenty-first century many (if not most) bogs and mires were not in pristine condition and what we see today are 'damaged bogs', that is, versions of the types described but altered to varying degrees by human exploitation and other activities. These impacts range from fairly moderate to verging on total destruction. Furthermore, many sites have indeed been eradicated.

Damage to sites generally involves drainage and sometimes associated peat removal or else ploughing and planting of trees. Desiccation of sites facilitates invasion by scrub and trees, and there is the decline or loss of sphagnum mosses and heathland plants such as *Calluna*. Further complications result from eutrophication either via rainfall deposition or from contaminated groundwater. With sphagnum mosses the impact may be a shift from peat-forming species like *S. magellanicum* and replacement by *S. fallax*, far less significant as a peat-former.

Temperate raised mires

Raised mires around the Baltic are described as '*Hochmoor*' or literally, 'high moor' or 'raised bog'. The British equivalents are more Atlantic-influenced, have flatter centres, and a less steeply sloping rand which drops gently to the surrounding mineral soil. There should be a surrounding 'lagg-fen' system but in most British sites which remain, this has been lost to agricultural 'improvement'. These bogs are distinguishable from blanket mires in having discrete margins and being generally set into a wider non-peat landscape. Formation often happens over shallow glacial or post-glacial basins or hollows. Across Britain and Europe these features were scattered across northern areas as the ice age glaciers retreated and the hollows created by the ice were frequently filled with impermeable boulder clay or 'till'. Raised above the surrounding land as the bog surface grows upwards, such mires have a form largely independent of subsurface topography.

The ecological successional processes that followed from the creation of a wet pool in a post-glacial hollow have been well-described from early-stage inception of the open water and fen. With open water in the basin there developed a surrounding fringe of fen vegetation, in and around the outer zone, carr of willow and alder. Materials in-washed from the surrounding land gradually combined with *in situ* generated sediments to infill the original lake, sometimes completely. It was once believed that the succession then proceeded to a stage of full forest cover but has since been disproved or at least shown to be rare.

However, as the basin filled with sediments and fen peat, the plants growing in the centre were increasingly deprived of in-washed nutrients and the local soil conditions became more acidic and infertile. The *Sphagnum* bog-mosses do well in such low-nutrient conditions and as they colonise the fen is gradually converted into a bog. Sphagnum mosses are interesting and unique as their leaves have small photosynthetic cells squeezed in various ways between large empty or 'hyaline' cells. These big 'empty' cells form effective water-conducting systems throughout the sphagnum plants. For the moss this system presents a large surface area of plant cells to the water and helps facilitate nutrient uptake from oligotrophic (infertile) water. The plants also release hydrogen ions in exchange for essential nutrients and these help cause further acidification of the bog-water. This process further favours sphagnum mosses which grow well in acidic conditions and so they tend to become dominant.

Sphagnum mosses grow from the very top or apex of the plant and the older stem below eventually senesces or dies. The dead material accumulates in a waterlogged zone below the plant and peat begins to form. The resulting structure conducts water rather poorly and so moisture is held within the peat layer and sphagnum mass rather than escaping. Over time the site becomes increasingly waterlogged. Furthermore, the mosses grow and as they extend upwards and peat builds up below, the mire rises to form a raised bog.

In most cases, these are lowland features and in Britain occur (or occurred) most widely in the north and west of England, in the Scottish Lowlands, and with outliers in Wales, and the east and south-west of England. They occur widely in Ireland too and historically were more abundant and widespread. The east English sites such as Thorne Moors in South Yorkshire have a strongly Baltic affinity. Most sites have been either entirely destroyed or at the very least severely modified by activities associated with agriculture and horticulture. However, a few examples remain in central Scotland, around the Solway Firth, and in Wales where the sites still have substantially 'natural' vegetation cover. A good number of other damaged sites are now in the beginnings of long-term ecological restoration. A major occurrence of British lowland raised mire is found in the remains of the once huge expanse of Flanders Moss in the carselands of the upper River Forth. Here, overlooked by Stirling Castle, there are still eighteen raised bogs that are reasonably intact.

The upward growth of raised bogs lifts their hydrology and associated available nutrient away from the surrounding landscape until the bog is totally dependent on rainwater for moisture, and importantly, for nutrients. In this state, the raised mire is described as 'ombrotrophic' or 'rain-fed' and is a highly nutrient-deficient ecosystem. It is this special quality that leads to the remarkable vegetation and associated fauna of these sites.

Blanket mires

The other major form of acidic peatland in temperate zones is the blanket mire or blanket bog. These areas are strongly associated with mild, wet climatic conditions such as the Atlantic-influenced zones of north-western Europe. In such circumstances the ground can be so waterlogged that dead plant materials do not decompose, and therefore accumulate as peat over the underlying mineral soil. This process is known as 'paludification' and in extreme climatic situations can occur over isolated exposures of rocks such as in Scotland on glacial boulders or erratics. More commonly, paludification proceeds over geological formations like boulder clays or till as found in the north Pennines of England and parts of Scotland, or else over acidic bedrocks like millstone grit in the English south Pennines and Peak District. In these cases, peat forms over soils subject to podzolisation associated with acidity, high rainfall, and low evapotranspiration rates. This eventually leads to the development of precipitation of a hard mineral layer called an 'iron pan' as dissolved iron oxide comes out of the solution to form a layer which is impermeable to water. The site becomes increasingly waterlogged and peat builds up. The initial stages are linked to fluctuating conditions of waterlogging and aeration as the water level varies and this triggers the formation of peaty-gleyed podzol soils. With more waterlogging, full peat formation may be initiated and the site begins to hold back more water because of the poor hydraulic conductivity of the peat. Ultimately, the peat may cover entire landscapes and in some areas accumulate to considerable depth.

Within the blanket peats there are 'fossil' remains of trees such as oak, birch, and pine and these mark periods of warmer, drier weather across the regions, such as during the Holocene Epoch (9,000 to 6,000 years ago). The tree cover in upland areas was probably quite patchy and relatively thin anyway, and these zones were favoured by early human settlers, from Mesolithic hunter-gatherers to Bronze Age and Iron Age farmers in the period from 5,000 to 2,000 years ago. In areas such as the English Peak District and Pennines, these human communities impacted on the vegetation and the soils. Shallow brown-earth soils developed under the vegetation cover were generally rather vulnerable skeletal soils rich in silica and with limited ability to hold nutrients. Combined processes of tree clearance, of burning associated with human settlement and particularly with tilling of the ground, caused rapid runoff and massive erosion of soils. This led to podzolisation and waterlogging and triggered development of blanket bog extending across wide areas of the landscape. The soils are seen today in the alluvial deposits of downstream lowland floodplains.

At the same time as these human impacts, there was a widespread deterioration of climatic conditions across these regions and weather became generally cooler and wetter. In at least some regions, the tree growth declined rapidly as soils became waterlogged.

As peat forms then the same basic processes occur as described earlier for raised mires. The sphagnum mosses colonise, hold water, and form peat and the process accelerates with acidification, low nutrient status, and waterlogging. This paludification can occur on sloping ground as long as climatic conditions allow. There are other processes which take place in the wider blanket mire environment, and these may link to topography and ecological successions. This may involve the infilling of open water sites and 'terrestrialisation' through what is known as hydroseral succession, ultimately leading to a raised bog, that is, domed peat and sphagnum areas within the blanket mire landscape. Along with these features there develops marsh and fen vegetation along watercourses which themselves become subsumed and enveloped into the wider peat deposits. The result is that so-called 'blanket bogs' or 'blanket mires' are made up of mosaics of peatlands of varying depths and different topographies. They include both ombrotrophic (or bog) elements and rheotrophic (or fen) components. The term 'blanket mire' covers this wider interpretation of the landscape which, whilst generally called blanket bog, is in practice a rather diverse ecological system.

The complexity of blanket mire systems can be considered in terms of topography, morphology, and hydrology. The report by Lindsay et al. (1995) provides an overview of the subunit or mesotypes based on hydro-morphological character:

1. **Watershed bogs:** these are ombrotrophic peat areas formed on the tops of hills
2. **Saddle Bogs:** these are in part rheotrophic mires which develop as infill within saddles or basins in the lower parts of the landscape and so receive water runoff from the surrounding slopes.
3. **Valley-side peats:** these sites are mostly with thinner peats, the depth of which varies with the slope. They tend to be less ombrotrophic due to flushing that influences the vegetation which may shift towards fewer sphagnum mosses and more fen species.
4. **Spur bogs:** In intermediate situation these spur bogs share characteristics of both saddle and valley-side features.
5. **Ladder fens:** This rather unusual type of mire is found in the Scottish Flow country and is similar to the 'ribbed mires' of eastern Canada.
6. **Minerotrophic fens:** These are areas of fen vegetation that develop in zones which receive water from the surrounding landscape. They may occur within blanket mire landscapes or else as separate entities, especially in lowland zones.

Fens

Whereas bogs are mires systems which develop in conditions of low available nutrients and input of water mostly or entirely from rainfall, fens receive water and nutrients from surrounding topographic areas. Essentially fens received water rather than releasing it. The resulting communities have higher nutrients and base-status (higher pH), and tend to be dominated by open water, reeds (*Phragmites communis*), and a diversity of higher plants like sedges, rushes, wetland grasses, and herbs such as meadowsweet (*Filipendula ulmaria*), common valerian (*Valeriana officinalis*), forget-me-not (*Mysotis* sp), and various orchid species. Established sites may follow successions to develop into wet woodland or 'carr' with willows (*Salix* sp), alder (*Alnus glutinosa*), and the like. The peat formed in fens is of a higher base-status and is dark and fibrous. As sites infill they become more wooded and drier, or else remain wet, but increasingly dominated by sphagnum mosses. In the latter situation, the mosses hold water and grow upwards until the surface of what is now becomes a bog, rises above the surrounding topography. The peat formed in the transfer to bog status is lighter coloured and spongier. The timescale for these changes, where they have occurred, is anything from hundreds to thousands of years and its progression can be tracked in the record of pollen grains laid down within the peat-bed. This can be carefully examined to reveal evidence of the vegetation changes which have happened over time on and immediately around the site. The progression from open water and reed-fen to raised bog is referred to as terrestrialisation and whilst the basic successional process as described by Tansley in the 1930s is clear and neat, the reality is not.

Although terrestrialisation is probably the better-known process, Wheeler & Shaw (1995) commented that in Britain paludification was probably more extensive. This is evidenced by widespread development of blanket peats across many upland regions and in low-lying northern zones too. The expansive peatlands of the East Anglian Fens also were formed more through paludification than terrestrialisation.

Terrestrialisation from open water and fen

The primary requirement for terrestrialisation to occur is that the water body is permanently wet, sufficient for the various 'seral' or successional stages of the hydrosere to take place. Furthermore, it is necessary that:

1. Plant species establish on the site
2. Water chemistry supports the growth of the early successional species
3. Environmental factors (e.g. light and temperature) support plant growth
4. Conditions at the base of the water body become anaerobic or anoxic and they allow dead, undecomposed, organic matter to build-up

So long as there are wet anaerobic conditions and submerged, dead vegetation accumulated, peat will continue forming. During the early stages of the succession, the peat-formation will be strongly influenced by mineral-enriched or base-rich groundwater and vegetation is described as 'fen'. As the peat, together with any in-washed materials from the surrounding landscape, fills the basin in which the wetland formed, the ecosystem reaches a critical point. From this key stage more peat can be laid down, but only if conditions remain waterlogged as the hydrology of the site is now independent of the source of mineral-rich groundwater.

For the developing peat to remain waterlogged (and anoxic), the site requires input of rainwater and sphagnum-dominated vegetation to absorb and hold the moisture. In some cases, the dead plant material, now above the surrounding water table, becomes too dry to be waterlogged, and so it decays and peat development ends. The terrestrialising system essentially attains steady-state equilibrium with minerotrophic (base-rich) fen or fen carr maintained by mineral groundwater influence or until the site progressively dries out. Tansley in 1939 described the now-classic hydrosere succession and suggested that peat growth can carry on through this steady-state condition. He also included the assumption that the site would be colonised by trees and shrubs and the weight of developing surface vegetation would press downwards to make the peat denser and squeeze it below the water table. This process described by Tansley would allow peat to accumulate but the successional change would eventually reach mature, dry, oak (*Quercus*) woodland. However, the Tansley 'climax' vegetation has never been convincingly observed in Britain; but then, the ecosystems present today have been hugely modified by human interventions. Wheeler (1980) indicated that the so-called climax would be fen woodland or as Moen (1999) described for Scandinavia, 'swamp woodlands'. This seems to be similar to the North American situation, and indeed, to my suggested historic vegetation over much of the English fens before drainage (Rotherham, 2014a).

However, if site conditions are favourable (i.e. wet) then a rainwater-fed wetland can establish and the peat surface may rise above the level constrained by groundwater. Such development moves the system from a minerotrophic fen to an ombrotrophic bog dependent on rainwater for moisture and nutrient inputs. In this situation, peat accumulates at around one millimetre of compressed peat per year and the surface gradually rises above the surrounding landscape. Eventually the peat mass starts to form into a dome with a rain-fed water table perched independently over the minerotrophic water table. When this occurs the ecosystem is ombrotrophic mire fed directly by rainwater and specifically called 'raised bog' or 'raised mire'. In northern temperate zones, the sphagnum mosses are the key to this bog formation.

In the southern hemisphere, with sphagnum present but less dominant, other species may take on the role of northern hemisphere sphagnum; for example, in Australia and New Zealand *Donatia* and *Empodisma* become important as long as the particular environmental conditions are suited to the development of raised mire. In particular, the necessary site hydrology and nutrient status are essential to the process.

In fully developed raised mire, there is a zone of fen vegetation surrounding the raised bog; and this is called 'the lagg' or 'lagg fen'. It is the interface between the rainfall-dependent bog hydrology and the groundwater of the surrounding landscape. Over time, as the ecosystem develops and grows, the bog edge and therefore the lagg may progressively move outwards until constrained, for example, by local topography. There may be additional terrestrialisation of open water in the lagg and paludification if the terrain allows this.

As with blanket mires, the process of paludification can take place and peatlands can form in the absence of a prior water body and associated terrestrialisation. This requires enough wetness (with high rainfall and low evapotranspiration) to allow peat-forming plants to colonise and establish. In such cases, peatland can develop directly onto mineral soil or rock surfaces.

4

TROPICAL AND SUBTROPICAL PEATLANDS

Tropical and subtropical peatlands

The issues of tropical peatlands and peat-forests are probably less familiar to many readers. However, this is a rapidly expanding field of study and in recent years there has been a big increase in awareness of these irreplaceable ecosystems and their roles in global climate moderation and in biodiversity conservation. The reader is directed to the extensive literature specifically on tropical peats by Page et al. (2002, 2006, 2009, 2011), and Page & Rieley (1998). Pearce (2017) also raises some important issues concerning tropical peat resources (Figure 4.1).

Eco-regional occurrence of tropical peats

Major areas include the following: Borneo peat-swamp forests (Brunei, Indonesia, and Malaysia); Eastern Congo swamp-forests (Central African Republic and Democratic Republic of the Congo); Peninsular Malaysian peat-swamp forests (Malaysia, Thailand); Ratargul Swamp Forest (Sylhet District, Bangladesh); Central and South America; and various smaller sites on tropical islands. The largest known area (56%) of tropical peatland and associated peat-carbon storage (77%) is in South East Asia. Here the peatlands are both extensive and thick with depths often over ten metres. The South American forests have the next most extensive tropical peatlands (24%) but they tend to be thinner and therefore contain proportionately less of the global tropical peatland carbon store. The other main terrestrial areas hold fewer peatlands. It is believed that the African forests have around 13% of the worldwide area and 8% of the carbon store, though the figures may vary as new sites are discovered. Central America and the Caribbean, mainland Asia, and Australia plus the Pacific islands hold only around 5%, 1%, and less than 1%, respectively, and overall have only about 3% of the global carbon store.

FIGURE 4.1 Indonesian peat-swamp forest by Wilter Malandi.

Tropical peatlands occur mainly in lowland areas in sub-coastal zones. They tend to be formed from the highly productive vegetation such as rainforest trees and associated higher plants. The forest ecosystems in the tropics are incredibly species-rich and diverse, and so there are significant regional differences in the plant species associated with peat and peatland formation. There are differences associated with altitudinal zonation too with plants that are lower-growing and more tolerant of low temperatures with increasing height above sea level. Page et al. (2011) have estimated the area of tropical peatland to be around 441,025 square kilometres or about 11% of the world's peatland resource. Estimates vary with a lower band of around 387,201 square kilometres to an upper band of 657,430 square kilometres or more, depending on specifically what sorts of peats and peat-soils are included. Furthermore, these estimates rely on current peatland area inventories and these mostly lack key data on matters like peat thickness and carbon storage. Rieley & Page note that the data leave much to be desired and the accuracy is variable from region to region, and from country to country.

So it is accepted that in some regions, the information on tropical peatlands is less than adequate. However, what we do know is that those tropical peatlands which remain are now being subjected to ever-intensifying changes in land use, and in particular, many are being converted to modern agricultural systems. These new landscapes include commercial plantations for products such as palm oil. The new systems often destroy biodiversity, including much that has been barely recorded before its loss, and they displace indigenous peoples, and haemorrhage carbon stored in peat back into the atmosphere. This process is well-advanced in South East Asia and particularly so in Indonesia and Malaysia. In these countries, most of the peatland area has already been deforested, drained, and converted to farming. The forest and peat frequently burnt as fire is utilised as a land-clearance tool. The new agricultural systems are often badged as 'sustainable' forestry which is a barefaced lie since alongside ecological damage there is major release of greenhouse gases (including carbon dioxide from the long-term storage in peat) and significant, contribution to ongoing climate change. Other regions and countries are somewhat slower than Malaysia and Indonesia in the destruction of their peatlands, but South America and in particular Brazil they are making major incursions into their remaining areas of peat-forest.

The issues confronting conservationists trying to protect and, in some cases, restore tropical peatlands are in effect a mirror that reflects the wider disparities of human condition and economics in the twenty-first century. Essentially, Western countries are relatively wealthy and those of the tropics (often described as 'Third World') are relatively poor. This means in effect that those countries with natural assets such as rainforest may seek to exploit it today without regard to the appalling impacts this will inevitably bring tomorrow. Furthermore, within these same countries there is generally a lack of effective democratic process and a massive gulf between rich and poor. Forests inhabited often by indigenous peoples frequently distanced from centralised and commercial decision-making processes are inherently disadvantaged. They and their peat-forests become the immediate casualties of longer-term social and environmental inequities.

An overview of tropical peat and peatlands

Tropical peat is a type of histosol soil (i.e. peaty soil with a deep surface layer of predominantly organic matter) and which is found in tropical latitudes. Rather than being composed of sphagnum moss as in temperate peats or reed-fens plants in fenlands, tropical peat is mainly dead organic material from trees. Organic matter content generally exceeds 75% of dry weight and these soils have a dry low bulk density around 0.2 mg/m^3.

The presently known areas of tropical peat are mainly in South East Asia (around 70% by area) along with Africa, Central and South America, and elsewhere around the Pacific Ocean. At a global level, tropical peatlands are important carbon sinks which store substantial quantities of carbon; their disruption or destruction therefore has serious implications for atmospheric carbon dioxide levels and hence potential climate change. Estimated areas and volumes of tropical peatland vary but could be something around 380,000 square kilometres of varying depths, but their precise occurrence is not fully known. However, what is clear is that tropical peatlands are currently under severe threat because of human as well as climate-induced change. The latter is itself in part precipitated by damage to the peatlands; a vicious circle.

So whilst tropical peatlands probably only cover around 0.25% of the global terrestrial area, they contain 50,000 to nearly 90,000 million tonnes of stored carbon or approximately 3% of the world's soil carbon resource. Other importance associated with these regions is that they support local indigenous peoples and richly diverse ecosystems including iconic and rare species such as the orangutan and the Sumatran tiger. The peat-swamp forests also produce valuable timber trees and other products of value to local communities including fruit, bark, resins, and latex. But these are contested territories and land-use changes plus fire (mostly linked to plantations such as palm oil), and logging that results in deforestation and drainage, are catastrophic. The impacts reduce carbon storage and contribute to greenhouse gas emissions.

Some problems may stem from a lack of awareness and understanding of how tropical peat-forests function and of their importance to regional ecosystems. These are incredibly fragile systems and vulnerable to any significant disruption. If the forest cover is lost and the peatland drained, then surface peat oxidises. If this happens then there is loss of stored carbon to the atmosphere and removal of floodwater buffering. The peat surface erodes and oxidises so land subsides and localised flooding may occur – climate change impacts also follow and coastal zone areas become more vulnerable to natural catastrophes such as tsunamis. Furthermore, this is an issue for temperate regions as well, the carbon release is not generally included in national audits and accounts for greenhouse gas emissions. This means that we consistently underestimate the rates of increase in atmospheric greenhouse gases, the extent of related human-induced climate change, and the contributions of individual nations to the problem. This is a crazy and illogical system.

Tropical peat-swamp forests have waterlogged soils that prevent dead leaves and wood from breaking down fully from decomposition. Research in Malaysia has shown that peat in tropical peat-swamp forest accumulates in extremely wet and hot conditions with sites waterlogged, nutrient-poor, anaerobic, and acidic. These conditions inhibit microbial activity that would otherwise cause breakdown. However, recent studies, also in tropical Malaysia, have shown how toxic chemicals in the leaves of some tree species are also important in preventing decomposition. These biotoxins have evolved in tropical forest trees as protection against insect herbivory, but a side effect is that they slow decomposition and help trigger peat accumulation.

The part-decomposed organic matter accumulates to create a layer of acidic peat of variable thickness. The peat-swamp forests are mostly surrounded by various types of lowland rainforest where soils are freer draining, and by brackish or saltwater mangroves in coastal zones.

Other tropical peatlands coexist with swamp-forests as part of the broader tropical and subtropical forest biome. Together these accumulate and store atmospheric carbon as soil organic matter and are among the largest near-surface reserves of terrestrial organic carbon on the planet. Importantly, these peatlands are very stable and this enhances their multiple roles such as buffering against climate change and as core sites for global biodiversity. Their ecological importance is unrivalled and yet they are some of the least studied and poorly understood ecosystems we have. They are also amongst the most threatened.

From the 1970s, peat-swamp forests have been felled, cleared, and drained at rapidly increasing rates and problems have been exacerbated by El Niño Southern Oscillation droughts and large-scale fires (mostly deliberate). A twist in this situation is that from being a major carbon sink the burning peat-forests are transformed into large-scale carbon dioxide sources. This mirrors the situation with temperate peat bogs and fens where, converted to arable farmland, they haemorrhage carbon to the atmosphere.

Countries with under-recorded peat deposits include Brazil, Peru, Venezuela, Colombia, Argentina, Colombia, China, India, and Bangladesh; and this is along with major areas in tropical Africa. In terms of the latter, research has identified new sites in Africa with up to ten times more peat than previously estimated. Simon Lewis and colleagues from the University of Leeds recently found a huge deposit of peat in the Cuvette Centrale region of the Republic of Congo. Once a massive lake, this area is the largest tropical peat-swamp in the world and now provides habitat for lowland gorillas (*Gorilla gorilla gorilla*) and forest elephants (*Loxodonta cyclotis*). The peat is several metres thick, formed from dead timber and other vegetation, accumulated in the stagnant water over around 10,000 years. The site extends over about 150,000 square kilometres and contains thirty billion tons of carbon. This carbon is equivalent to that in all the trees of the entire Congo basin rainforest and equal to twenty years of carbon dioxide output from the USA.

Like the African peatlands, those of South America are turning out to be far more significant than previously believed. In the Peruvian rainforests of the upper Amazon basin in north-eastern Peru for example, a team from the University of St Andrews found a large, previously unmapped peatland extending over 35,000 square kilometres. Known as the Pastaza-Marañón Swamp it was described by researcher coauthor Katy Roucoux as, "the most carbon-dense landscape in Amazonia" (Roucoux et al., 2013). In only about 3% of the land area of forested Peru, this holds three billion tons of carbon which is about half as much of the country's entire forest cover.

According to Pearce (2017), this seems to be an emerging pattern with Thomas Gumbricht (scientific consultant at the Center for International Forestry Research in Bogor, Indonesia) (Gumbricht et al., 2017) reporting that tropical peatlands probably cover more than three times the area currently calculated. This would make it 1.7 million square kilometres rather than 440,000 square kilometres, but also, with volume perhaps fourfold higher, the volume would amount to 7,000 cubic kilometres. The conclusion would then be that global tropical peats hold 350 billion tons of carbon rather than the 88 billion tons presently assumed. This would amount to over 200 years of USA greenhouse gas emissions. It should be noted, however, that there are concerns about the methodologies employed in some of the recent studies that may have resulted in overestimates. Either way, it seems reasonable to assume that there is more tropical peat than previously realised and this may amount to a vital element in combating global climate change.

Development of tropical peatlands

Tropical peatlands occur in three main regions, Central and South America, Africa, and South East Asia. Around 62% of the world's tropical peatlands are found in the Indo-Malaya region (80% Indonesia, 11% Malaysia, 6% Papua New Guinea, and smaller areas in Brunei, Vietnam, the Philippines, and Thailand). Indonesian peatlands are found on the three islands of Sumatra (8.3 million ha), Kalimantan (6.3 million ha), and Papua (4.6 million ha). These peats developed in low-lying areas like deltas, floodplains, or shallow lakes such as meander oxbows of great rivers. Development seems to follow general hydrosere successional stages with eutrophic open water colonised by water plants and transformed into a waterlogged swamp with grasses and shrubs. This gradually morphs into peat-forest and the peat continues growing and accumulating to the point that in some areas a raised, convex-shaped dome is formed. Some peat may be pushed out to the dome fringes or zones in-between the core accumulations. The peat growth may rise to four metres in coastal peats and to eighteen metres in inland sites.

With tropical peat formation, the early stages are mostly topogenous or minerotrophic as they receive nutrients from rivers or groundwaters. However, like with temperate peats, as the deposit deepens and the peat-dome elevates above the surrounding water table, they become ombrotrophic. Water input is then exclusively from rainfall and so this becomes a low-nutrient system. With calcium particularly in short supply, the peat is highly acidic. At this stage, biodiversity may be reduced and any tree growth is stunted.

Inland and coastal peats differ greatly in age with many coastal peat sites beginning to form during the mid-Holocene Epoch (around 8,000 years ago) whereas some inland peatlands are older and began in the Late Pleistocene Age (*c.* 26,000 years ago). In coastal zones, the initiation of peat formation was strongly influenced by a rise in the sea level which in turn was affected by factors and events like the El Nino being less intense and by glacial-related changes which triggered alternating periods of inundation and of drying up of the seas. As ice sheets thawed, the coastline moved inland as the sea level rose. The oldest coastal peats in tropical Asia today are about 8,000 years old.

Rates of growth over long periods of time in the inland peatlands were sometimes very slow and varied from a few millimetres per year to maybe only 0.05 millimetres. The inland sites are driven mostly by climate and some topographic influences but little or no effect of any rising of the sea level. Sites are located about twelve to twenty metres above modern sea level, and historically this has not risen more than about six metres in the last 100,000 years.

Examples of tropical peat-forest and issues in South East Asia

Indonesia

Indonesia has about 50% of the world's tropical peat-swamp forest which were formerly major components of ecosystems there with 16.5 to 27 million hectares. In their original state, Indonesian peat-swamp forests were carbon-sequestering systems. However, these forests have been lost to deforestation felling, drainage, and conversion to agricultural lands and other functions. The reduction has compromised the ecosystem service functions

Over recent decades, the Indonesian government has pursued the 'Mega Rice Project (MRP)', whereby it has sanctioned the draining of over one million hectares of Borneo peat-swamp forest which was converted to agriculture. Indeed, between 1996 and 1998, over 4,000 kilometres of drainage and irrigation channels were cut. At the same time, deforestation rapidly accelerated through both legal and illegal logging and also burning to remove the trees. Infrastructure such as roads, railways, and water channels for legal forestry, also facilitated the areas to become accessible for illegal cutting and the government did little or nothing to prevent this.

Across the MRP region, the forest cover fell from 65% in 1991 to 46% in 2000, with ongoing clearance since that time. Assessments suggest that virtually all sellable timber trees have been taken from within this area. The intention had been that the cut channels would irrigate the region whereas in practice they drained it and today where the forest floor once stood up to two metres deep in water, it is bone dry. Late in the day, and after considerable damage, the Indonesian government halted the MRP. However, the region is now dominated by huge numbers of fires from burning peat and burning trees and the lowlands north-west and south-east of the region are shrouded by thick, grey smoke. In many areas, this adverse environmental impact has seriously affected a potentially lucrative tourism economy for the country.

The wider environmental impacts are even more drastic with research for the European Space Agency finding 2.57 billion tons of carbon released to the atmosphere in 1997 alone because of burning peat and associated vegetation in Indonesia. The estimate suggests an amount close to 40% of the mean annual global carbon emissions from fossil fuels. This is a major factor in the greatest annual increase in atmospheric carbon dioxide concentration since data records started in 1957. Catastrophic fires in 2002 to 2003, further released 200 million to one billion tons of carbon into the atmosphere, and these were associated with government corruption and illegal business activities. The effect on tourism across a wide area of South East Asia was devastating as palls of thick smoke clouded

FIGURE 4.2 Indonesian peat-swamp forest by Wilter Malandi showing a huge tropical tree.

a wide region internationally. A further consequence now is that from being a centre for carbon storage worldwide, Indonesia is today the world's third-largest carbon emitter. This is mostly due to the felling, draining, and burning of its ancient peat-swamp forests.

A key issue for conservationists and scientists is to find ways to highlight the significance of the forest resources and their peat. For peat-swamp forests there need to be alternative ways to make these areas productive for local people, including indigenous communities, but in ways which are environmentally sound and sustainable (Figure 4.2).

Peat-swamp forest in Kalimantan

Kalimantan is the expansive Indonesian part of Borneo and has extensive peat-swamp forests, with huge trees up to seventy metres high. These are totally different landscapes from the familiar peatlands of northern temperate and boreal areas where sphagnum mosses, grasses, sedges, and shrubs dominate. The anaerobic, seasonally flooded peat-beds can be up to twenty metres deep with a spongy, unstable, waterlogged, substrate that is highly acidic (pH 2.9 to 4.0) and nutrient-poor. The water in the forest is dark brown because of tannins leached from fallen leaves and the peat, and these are also called 'black-water swamps'. Even in the dry season the peat is waterlogged with pools of water standing amongst the trees. The actual water level on the peat is generally around twenty centimetres under the surface, but this may fall to about forty centimetres under exceptionally dry periods like El Nino conditions. This lowering of the water table makes the sites vulnerable to deliberate and accidental fires.

The peat-forest histosol soils have a high carbon content with typically 70% to 99% organic material. In temperate peatlands this high carbon content is maintained by wetness and low temperatures, whereas in tropical peatlands it is stabilised by waterlogging. Therefore, any changes to water status here can trigger major carbon release into the atmosphere.

Conservation and preservation

Conservation of tropical peat-swamp forest has had limited success and the efforts pale when compared to the widespread devastation of commercial logging and deforestation. There are attempts by conservation bodies like 'Borneo Orangutan Survival' for example, to preserve the peat-swamp forest of Mawas using carbon finance and debt-for-nature-swap. Habitat disturbance from deforestation has dramatic impacts on orangutan populations in mixed swamp-forest. The presence of a large, self-sustaining orangutan population in Kalimantan's peat-swamp forests, for example, emphasises the need for greater and effective measures. Furthermore, the orangutans are merely the iconic biodiversity headline species in forests which have superabundant endemic species including large numbers as yet unknown to science. As the forests burn we are losing species we don't even know we have. Currently, around 6% of the original peat-forest is in protected landscapes, the largest being Tanjung Puting and Sabangau National Parks. At the same time, logging is on-going and intensifying elsewhere.

Palm oil agribusiness is the main driver for East Asian deforestation along with illegal logging in places like South Sumatra. Surveys indicate that complete removal will have occurred by 2030. Projects initiated under concepts such as 'reducing emissions from deforestation and degradation' (REDD) address deforestation as a global issue. Destruction impacts on biodiversity and livelihoods of forest-dependent communities but is also a major climate change driver which leads to around 20% of global greenhouse gas releases. REDD projects are intended to address forest destruction and to provide protection from agricultural encroachment. The idea is that such initiatives should benefit biodiversity and improve environmental quality for surrounding communities.

Organizations like Wetlands International are now working with the Indonesian government to provide improved policy and spatial planning guidance on these issues. The intention is to deal with issues of the unsustainable palm oil expansion in Indonesian peatlands and the destruction of mangrove swamps. Perhaps, rather controversially, the projects engage with palm oil industrialists who are at the heart of this problem. The idea, however, is to promote 'best management practices' for tropical peat-swamp forests and engage local communities who often are unaware of key aspects of natural resource management. Work in the field tries to support local people in restoring mangroves and peatlands.

Examples of tropical peat-forest and issues in South and Central America

Brazil, Peru, and the Amazon

There are various types or categories of tropical peat-forests depending on climate and topography. In lowland river floodplains and swampy areas, there are extensive tropical rainforests. On higher ground and mountain slopes, cloud forests and elfin woodlands are found in the humid tropics. Along with rainfall these areas receive added humidity through the capture or condensation of water droplets. Virtually undecomposed organic matter may accumulate as peat to depths of over four metres. There is also paramos vegetation which occurs in tropical ecosystems between upper zone of continuous, closed-canopy forest and the higher limit for higher plant growth between 3,000 metres and 5,000 metres, known as 'zacatonales' (in Mexico and Guatemala), 'páramo' (in Central and northern South America), 'jalca' (in northern Peru), and 'puna' (in drier parts of the central Andes). Swampy cushion mires or 'turberas' occur commonly in this vegetation, particularly in the higher level grass páramo zones.

Using new methods to assess and analyse terrestrial vegetation it has now been suggested that South America may hold 46% of the world's total of tropical peat and perhaps more than Asia (which was formerly considered the largest reservoir of tropical peat). Newly discovered peat areas in Brazil are relatively shallow Amazon basin deposits along the Rio Negro and Rio Branco. The Amazon is certainly a major reservoir of tropical peat and Brazil rather than Indonesia may be the world leader at least in tropical peatland area if not peat volume. Droughts and illegal burning of peat-forests are impacting on peat deposits and this may have greater implications for greenhouse gas release than previously estimated. El Niño droughts are further exacerbating the situation and more recently Brazilian governmental involvement in the ethnic cleansing of indigenous peoples and deliberate firing of forests has been catastrophic.

Somewhat overlooked is that Brazil has around 612,000 hectares, (about 0.07% of the county's area), of acidic peatlands. These are in southern Brazil in Paraná state in the high altitude grasslands which extend over 28,000 square kilometres and the bogs have herbs and bryophytes with a dominant cover of sphagnum mosses. Here these high-level (1,000-metre) sphagnum peat bogs function as reservoirs of carbon and also filter and regulate water flow into some of the major hydrographic basins of southern Brazil.

Conservation issues

As with countries like Indonesia, the main issues in and around the Amazon are deforestation and agriculture. Again, palm oil production is expanding rapidly in South America; by 2015, 6% of global output was coming from the region. As in other areas, these are strongly contested territories and local indigenous peoples are being ruthlessly removed. Other problems here and in Asia include rice fields, sago palm plantations, and other agriculture. There are also increasingly widespread impacts of mining for minerals.

The subtropical peatlands

Often neglected in discussions on peatlands are those in the subtropical zones which include the southern USA, some areas of North Africa, the Mediterranean, the Middle East and even parts of Australia. The main example described here is that of the Florida Everglades in the southern USA (Grunwald, 2006; McVoy et al., 2011; Hohner & Dreschel, 2015; van der Valk et al., 2015; Dreschel et al., 2018).

The Everglades is on limestone bedrock relatively impermeable to water and this is along with a coastal ridge forming a barrier in the east. Together these caused the retention of water in the Everglades basin and combined with the warm, wet climate suit the lush growth of herbaceous vegetation and deposition of organic peat soils. These grew at around seven centimetres per century initially but rising to twelve centimetres per century in the last 1,000 years. Before the nineteenth century, peat accumulated enough to create a dam along the south shore of the Lake allowing water levels to increase and overflow southwards. This meant water was continually flooding the south Florida peninsula, and especially so in wet seasons. Peat accumulation provided enough water holding capacity to retain moisture during drier seasons and favour peatlands.

The Everglades region has four major and distinctive peat deposits associated with it. These types are:

1. Everglades peat which is most abundant and formed from partially decomposed sawgrass leaves and roots in ridges within the ridge and slough landscape,
2. Loxahatchee peat is from aquatic plants like water lilies in slough bottoms,
3. Okeechobee muck, and
4. Okeelanta peaty muck.

The latter two types of peat are formed from overwash during periods of high water levels in the lake and have more inorganic material (around 10% dry weight). It is worth noting that a part of the human impact associated with agriculture has been peat exploitation here. Both the two mucks and the Everglades peat have been widely cut and used for farming 'land improvement'.

The Florida Everglades is a major area of patterned peatland formed from sawgrass or sawsedge (*Cladium jamaicense*) and other aquatic and swamp plant material. This has accumulated over thousands of years with initiation in some cases around 5,000 years ago, and much active growth over at least 2,000 to 3,000 years. Peat formation was due to attenuated plant decomposition because of inundation and waterlogging. Surveys in the late 1800s, identified deep, three metres or more, organic peat soils. Major efforts to facilitate economic development of what at the time was considered a wasted region focused on drainage to allow agricultural development. The main objective was to lower the water level of Lake Okeechobee and drain the lands around it. This scheme successfully lowered the wider Everglades water table and triggered agriculture across around 2.6000 square kilometres of sawgrass swamp south of the lake, an area then named the 'Everglades Agricultural Area' (EAA). To this day it is dominated by sugarcane farms.

So the peatland was drained in the late 1800s for a mix of agriculture and urban development, in effect to wrest a wild wet, wasteland from nature and turn it to productive civilisation. Major drainage began with the digging of four 'muck canals' which by 1917 carried water from the lake southwards to the east coast. Much of the region is now criss-crossed by canals and levées. Recent research has indicated that around half of the original land surface area has been lost from the peatland. So the Everglades patterned peatland, which formerly extended over around 11,000 square kilometres of South Florida (McVoy et al., 2011), is now about 5,500 square kilometres. Also, following drainage, begun in the late 1800s, the organic peats have shrunk and the land surface across much of the agricultural area has subsided from up to 1.7 metres. Fire and oxidation (bacterial and chemical), and of course wind-blow erosion, have led to losses of around two and a half centimetres per year. This loss equates to around ten billion cubic metres of peat dissipated to the atmosphere and the ocean. Carbon emissions from these Everglade peat-soils emissions have been measured as between 0.4 and 2.67 $g/m^2/h$, and Dreschel et al. (2018) suggested that 1.3 billion metric tons of carbon dioxide had been given off by the Everglades peatlands since the pre-drainage condition in about 1880. This colossal emission due to peat loss equates to about a quarter of the carbon dioxide emitted by the entire USA in 2015 (which was 5.2 billion metric tons). The emission of peat-based carbon from the Everglades has made a significant contribution to global atmospheric carbon.

Hohner & Dreschel (2015) used historical and recent datasets, spatially referenced soil bulk density values, and carbon contents to calculate peat-mass and carbon-mass to compare the region's pre-drainage and present-day peat volumes. Their figures are hugely informative and suggest as a broad-brush estimate, that the Everglades today holds under 24% of the original peat volume, 17% of the mass, and 19% of the carbon. In other words, 81% of the original carbon store has been released to the atmosphere as an active greenhouse gas. Dreschel et al. (2018) then assessed original peat volumes and remaining peat volumes across the Everglades and the amount of peat-loss over around 150 years. They estimated the loss of peat-mass and carbon to be 900 million metric tonnes for the former and 300 million metric tonnes for the latter. This peat loss has major implications for future hydrological, ecological, and landscape restoration or recovery for the region.

The 1800s drainage and conversion to farming with intensification in the early 1900s triggered major environmental problems throughout the wider area. These came to a head following a series of dry summers which caused in major peat-fires across the Everglades. In wet years agricultural land and towns south of the lake were badly flooded. Furthermore, the disruption of soil hydrology caused saltwater to creep underground, contaminating water supplies to both the emerging townships and agriculture. The immediate problems led to the 1948 Central and Southern Florida Project for Flood Control and Other Purposes (C&SF Project) with 2,300 kilometres of levées (man-made embankments of soil and rock), canals, and associated structures enclosing large parts of the region to form Water Conservation Areas (WCAs). These were shallow reservoirs to control water-levels so water might be held or released as needed. A watershed moment was the establishment in 1947 of the Everglades National Park (ENP) to protect the southern zone of the Everglades. Together these initiatives created the present landscape of the Everglades Protection Area (EPA) (combining the WCAs and the National Park). Planning for the region now includes the Central Everglades Planning Project which intends to speed up recovery of Everglades peatlands by taking more water through the area. The idea is to create increased storage, treatment, and movement of water in the areas south of Lake Okeechobee. The works involve a mix of removal of some canals and levées in the core Everglades and the blocking of others to retain water in the National Park. At the same time, the works protect from flooding the urban and farming lands situated in the east. Hopefully increased water within the central Everglades will halt additional peat loss from oxidation and help trigger the deposition of new peat to replace some of the losses since the drainage.

The Everglades peatlands are an example of the way that these ecosystems can be strongly disputed spaces with agriculture and urbanisation triggering massive loss and degradation. Similar trends are clear on the tropical peat-swamps such as in Indonesia. Ultimately, but frequently very late in the day, the impacts of 'improvement' and drainage are clearly not sustainable, and the effects on local indigenous peoples are devastating. Furthermore, alongside catastrophic declines in associated biodiversity these areas switch from carbon sinks to carbon emitters on a global scale.

Subtropical peats occur elsewhere around the world with moderate accumulations in countries like New Zealand and even Australia (e.g. Adam, 1987, 1992).

In the latter case, the peats are in coastal mangrove swamps and similar areas with around 11,500 square kilometres of mangroves, mostly found on the northern and eastern coasts in the intertidal zones of tropical, subtropical, and some temperate coastal rivers, estuaries, bays, and marine shorelines – Queensland has 44% of Australia's total, Northern Territory 37%, and Western Australia 17%. However, less than 1% of Australia's forest cover is mangrove. Whilst sphagnum peat is rare in the continent, it does occur in the damp areas of Tasmania and the Australian Alps. In New Zealand, sphagnum is also quite uncommon and bogs are mostly of rushes (family Restionaceae). These bogs look and feel different from those of northern temperate zones. New Zealand also has mangrove swamps but they only occur in the north of North Island, extending naturally on the east coast as far south as Ohiwa Harbour in the Bay of Plenty. Along the west coast they reach their natural limit at Raglan Harbour though they are occasionally found a bit further south in Aotea and Kawhia harbours. Whilst mangroves mainly occur in warmer climates and achieve greatest diversity and size in tropical zones, where conditions are cooler, such as New Zealand and southern Australia, there is only one species *Avicennia marina* (also known as *Avicennia resinifera*).

5

PEATLAND ECOSYSTEM SERVICES

An introduction ecosystem service delivery

Bonn et al. (2010, 2014, and 2016 for example) have developed ideas and outputs from the 'Defra Ecosystem Approach' framework (TEEB, 2009, 2010 a, b, 2011a, b, c; Dunn, 2012), but specific to peatland systems. This approach has been informed by that of the Millennium Ecosystem Assessment (Millennium Ecosystem Assessment, 2005), and is today taken as the standard way to assess the contributions to humanity of natural ecosystem functions. The approach helps those such as planners and economists to take ecological aspects of the wider environment into account when deciding on utilisation and on conservation or restoration interventions.

Services delivered

Given a growing understanding of the ecology and hydrological functioning of peatlands such as bog, mires, wet moors, and fens (e.g. IUCN, 2011), it is increasingly clear that they deliver massive 'ecosystem services', such as water management and flood mitigation, biodiversity, climate change resilience, and more (Haines-Young & Potschin, 2008). In recent years therefore, there have been moves to better understand the ways in which such services are delivered, and how peatland losses and degradation have compromised these (e.g. Watson & Albon, 2011a, b). Such a standpoint has been a platform from which restoration and reconstruction of functioning peat landscapes can proceed. This chapter presents current thinking and examples of the ecosystem services approach, and the global application of these emerging concepts.

Understanding the critical roles of peatlands in future delivery of ecosystem services and importantly in climate change limitation and mitigation, has been central to recently emerging conservation initiatives. First, there is the need to have good insight into problems and processes, and especially into the scale of damage to peatlands and the implications of that on vital services such as water management and carbon storage. The second aspect of this is the development of effective economic markets to place values on those ecosystem services and to facilitate 'trading' in things like carbon offsets. Of course, a counterbalance to these economic tools is that society and its governments simply need to learn that we must halt the damage and do what we can to restore ecosystem functioning. This may require payment through the public purse, but failure to do so will ultimately cost far more; a genuinely inconvenient truth and unpalatable to current political thinking.

With robust recognition now at national and global levels of the benefits associated with functioning peatlands, there are widespread moves to slow the decline and to (where possible) repair the damage. Furthermore, in recent years there have been great steps taken to develop innovative techniques to restore functioning peatlands, or at least to re-establish the key ecological and hydrological processes that may in time lead to restored peat ecosystem services. Some recovery can be expected very quickly but in other cases we are talking about a long game of decades or even centuries. A bog which took maybe a decade to destroy may have taken 500 to 1,000 years or more to develop; so, recovery is not instantaneous.

Such long-term commitments to reinstatement require combinations of both public and private funding and these must operate at both national and international levels. The first steps in this process have been through the emergence of new funding sources from voluntary and compliance carbon markets in relation to climate change mitigation. However, to be fully effective these processes need to expand beyond the voluntary and to be underpinned by robust methodologies and science in order to quantify benefits against financial outlays. The reality of carbon trading can be rather complicated and not necessarily a precise tool. Furthermore, the repair to a functioning state of a damaged peatland becomes more expensive, more difficult, and with greater risk of failure the more seriously damaged the site. There are other ethical considerations too whereby on an English, temperate blanket mire, for example, a landowner may have been subsidised to drain the site and to overgraze it, and now receives funds to put it back into good order. Another landowner who maintained a site in undrained condition and did not utilise grant-payments to overgraze the area may be disadvantaged again unless he or she is paid simply to maintain the status quo. This is a fundamental problem that needs to be addressed.

Additionally, in order to facilitate this step-change in attitudes to peatlands, there is the need for international agreements and protocols, and for national policies supported by appropriate legislation. In this context, peatlands have been acknowledged as priority habitats for action under various international treaties such as the 1992 UN Convention on Biological Diversity (CBD) and the associated Nagoya Protocol, and the 1992 UN Framework Convention on Climate Change (UNFCCC) and its associated 1997 Kyoto Protocol. The latter commits state parties to reduce greenhouse gas emissions based on the scientific consensus that (Part 1) global warming is occurring and (Part 2) it is extremely likely that human-made carbon dioxide emissions have predominantly caused it. These conventions recognise peatlands and their restoration as vital to the attainment of targets for both biodiversity and the climate. However, it must be noted that the already accelerating rate of climate change currently experienced is speeding the decline of currently damaged peatland systems and thus making their recovery more difficult and considerably more expensive. These changes also make reinstatement even more urgent.

Peat ecosystem services

The report to Defra on 'Ecosystem services of peat' Phase 1 (Bonn et al., 2010) drew together much existing knowledge on ecological services associated with mire systems. In this context, peatlands have been recognised as providing crucially important ecosystem services (Bonn et al., 2009). The report argues that peatlands are of national and international importance for provision of food and fibre, water supply, climate regulation, and maintenance of biodiversity. Furthermore, they provide opportunities for recreation, inspiration, and cultural heritage. The latter of course should include archaeological and palaeo-ecological remains which can provide major insights into landscape change and climate change; though in practice these are rarely considered.

Many of these areas are of exceptional natural beauty, provide centres of species distinctiveness and richness, and historically have been extensively managed for food production, game, fish, and also building materials and fuel extraction. Whilst bogs may be inherently of low productivity, fens, on the other hand, can be very productive. Indeed, they were highly valued by local commoners during medieval times and up to the nineteenth century.

These ecosystem service values are reflected in the fact that most UK peatlands have national and international conservation designations. As noted earlier, global peatlands cover over 400 million ha (Bather & Miller, 1991) and in the UK, peat extends over 1.58 million ha or about 7% of the land area, though as noted this is probably significantly underestimated due to historic land-use change and peat extraction. The Bonn et al. (2010) report addressed the main ecosystem services provided by peatlands as defined based on the commonly used framework of the Millennium Ecosystem Assessment (MEA, 2005) and the Defra Total Economic Value (TEV) (see Dunn, 2012) approach which identify four major categories of ecosystem services that directly affect human well-being. Accordingly, the main peatland services are:

1. Provisioning services: Associated with products from ecosystems, such as food, fibre, freshwater, and sources of energy or building materials. Today blanket bog peats are used for rough grazing, while fen peats, being more fertile, are often converted to agriculture; historically they too were grazed.

A contemporary product of upland peatlands is their role in water supply with upland catchments providing over 70% of Britain's freshwater.

Peat extraction is destructive to mire ecosystems but milled peat (particularly from lowland sites) provides various services, the most common being a soil improver and growing medium for horticultural services. Peatlands may also provide sources of renewable energies such as wind power, small hydroelectric schemes, and wood fuel extraction, though these are often destructive to the peat sites. Peatlands may also provide locations for major transport routes, though again, this is hardly a 'peatland' service (Figure 5.1).

2. Regulating services: The regulation of natural processes, including climate regulation, air quality regulation, water purification, and natural hazard regulation are further examples of peatland ecosystem services. It is calculated that the UK's peatlands store over 5,000 million tonnes of soil carbon. According to Milne & Brown (1997), this considerably exceeds the total carbon stored in the UK's woodland vegetation of around 92 million tonnes. The Defra 'Ecosystem Services of Peat – Phase 2' report notes that peatlands exhibit a tight balance of greenhouse gas sequestration and emission depending on their particular conditions (Baird et al., 2009). Peatlands are important in water regulation, affecting water quality and flood regulation including mitigation of natural hazards such as flooding and wildfires.
3. Cultural services: Provision of non-material benefits from ecosystems are those such as recreation opportunities, enjoyment of landscape aesthetics, biodiversity and cultural heritage, along with spiritual enrichment, and education. In this context, many peatlands are the most extensive areas of intact and unbroken landscapes and could be major tourist attractions for

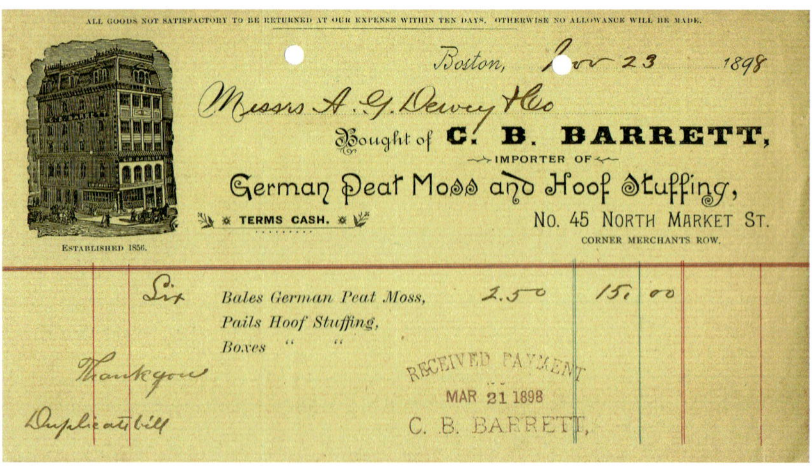

ALL GOODS NOT SATISFACTORY TO BE RETURNED AT OUR EXPENSE WITHIN TEN DAYS. OTHERWISE NO ALLOWANCE WILL BE MADE.

Boston, Nov 23 1898

Messrs A. G. Dewey & Co

Bought of **C. B. BARRETT,**

IMPORTER OF

German Peat Moss and Hoof Stuffing,

TERMS CASH.

No. 45 North Market St.

CORNER MERCHANTS ROW.

ESTABLISHED 1856.

Six Bales German Peat Moss, 2.50 15.00

Pails Hoof Stuffing,

Boxes " "

Thank you

Duplicate bill

RECEIVED PAYMENT

MAR 21 1898

C. B. BARRETT,

FIGURE 5.1 Bill of sale of German peat moss and hoof stuffing, Boston, USA.

countryside recreation. Such opportunities also bring major economic benefits to often geographically remote areas, and unlike destructive extraction, are sustainable on a long-term basis.

4. Supporting services: these are services necessary for the other ecosystem services to function. They include functions such as nutrient cycling, microclimate regulation, soil formation, and photosynthesis

UK Peatlands and their services

Provisioning

With naturally low agricultural productivity (due to soils, waterlogging, and often access, topography, and climatic conditions), peatlands are mostly classed as poor quality farming land with severe limitations. This restricts use to permanent pasture or rough grazing (Agricultural Land Classification, MAFF, 1988) and this is especially so for upland sites. With lowland peatlands drainage, warping, and lime and fertiliser treatment have often increased their short-term productivity to high levels, but loss of the organic component each year means this may only last for another forty years or so in for example, England.

The different potentialities of upland and lowland peatlands meant that most upland sites have seen moves to either marginal pastures or overgrazed unenclosed lands, whilst lowland fens become 'breadbasket' arable. In the last fifty years or so, these changes have been driven by massive public subsidies for agricultural improvement.

With large-scale upland drainage schemes stock densities increased up to 400% triggering major overgrazing and degradation. More recent de-coupling of agricultural subsidies from production but with links to environmental benefits has helped trigger a degree of recovery in some upland areas. Payment for ecosystem service delivery is helping with this change and in the case of peatlands there may be direct economic benefits to commercial providers such as water utilities. Intact peat bogs help remediate the discolouration of drinking water and save on otherwise expensive removal downstream. This is one of the most significant examples of where there can be direct, commercial, economic payment for an ecosystem service recovery.

In lowland peatlands, food provision services are more complicated than in the uplands, especially when sites have been drained and 'improved'. In areas such as the Somerset Levels and Moors, the grazing of dairy and beef cattle tends to dominate. However, up to the nineteenth century, the Levels were mostly used for summer grazing as the region was inundated by winter flooding. Taking Somerset as an example of a widespread phenomenon in the English fenlands, to intensify twentieth-century agriculture, after the Second World War food crises, large pumping stations (such as Gold Corner), were constructed. These were able to lower water levels via man-made drains like the South Drain, and so providing accessibility for machinery the whole year round (Williams, 1970). It was this

FIGURE 5.2 Pat O'Shea, Irish bog oak dealer with his wares County Cork, Ireland.

peak in the progressive desiccation of these wetlands that altered their ecological and landscape character so dramatically. However, in many cases this was the culmination of processes extending back over centuries.

Both lowland and upland peatlands have been widely drained and planted with timber crops. This provides a source of timber and fibre but is very detrimental to other services. Wood was harvested sustainably from undrained fens until the twentieth century. Contemporary services include biomass energy crops but again, in a peatland landscape these are very damaging. 'Sustainable' energy provision may include wind turbines located in and around peat landscapes. Once again there are issues of landscape sensitivity and of disruption to peatlands that results in carbon release from disturbed or desiccated peat. Another obvious energy source from peatlands has been peat fuel harvesting, which is hugely detrimental and unsustainable. Already noted, peatlands, particularly in upland areas are very important in the provision of fresh water (Figure 5.2).

Regulating

Carbon sinks and carbon flux

Taking the UK as an example, peatlands form the biggest single carbon reserve with about three billion tonnes of carbon stored. This is more than the combined total for the forests and woods of England and France. By comparison, the UK woodland and forest estate hold approximately 150 million tonnes of stored carbon and this is relatively shorter-term storage than that in deep peats. Also importantly, is that about half of the UK's soil carbon is stored in only 8% of the country's terrestrial land area.

Climate regulation

Taken as a part of the terrestrial biosphere, the northern hemisphere peatlands are the single most important terrestrial carbon store. They are a globally important sink for potential greenhouse gases. Gorham (1991) estimated that 20% to 30% of the global terrestrial carbon is locked away in only 3% of its land area. This northern hemisphere peat-based carbon store is calculated to be around 4.5 Gt C. Over the time of the Holocene, such northern peatlands have stored atmospheric carbon at about 0.96 Mt C per year. From this perspective, the peat ecosystem is both a substantial long-term store and also a vast potential sink for atmospheric carbon.

Alongside carbon-related issues, peatlands can also store nitrogen and with the wet state of such environment peat soils may be a source of nitrous oxide which is a powerful non-carbon greenhouse gas. In peat, most nitrogen is derived from plant matter and stored in organic form, but there is additional nitrogen from atmospheric fallout. Peatlands release small amounts of nitrous oxide from microbial decomposition and so the peat can both hold nitrogen and release nitrous oxide back to the atmosphere.

Taking the northern hemisphere peatlands as a whole, many areas have experienced land management which has damaged carbon storage capability and has significantly reduced climatic regulation functions (Holden et al., 2007). An example of such practice is the use of drains to enhance agricultural productivity (Holden et al., 2004) and this has affected upland bogs, lowland bogs, and lowland fens.

Overall, land management practice offers threats and opportunities in terms of peat-soil carbon budgets. Historically the threats have prevailed and indeed still do in many intensive arable zones. Site management for farming or forestry may damage peat and reduce the carbon sink and transform the peat soils to net sources of atmospheric carbon. Nevertheless, with current awareness and knowledge of peatland functioning, land management can be turned to create opportunities to enhance carbon uptake and long-term storage. After major damage, full restoration of a functioning peat bog system can take a very long time. However, recent research has demonstrated that benefits start to accrue immediately following straightforward actions such as re-wetting.

Ecosystem services in tropical peat-forest

Less research has been undertaken into the ecosystem service functions of tropical peatlands, and sadly, often for political and socio-economic reasons, these areas are currently under severe threat. However, it is well-known that tropical

FIGURE 5.3 Sami turf dwelling in Lapland early 1900s.

rainforests are a key part of the global ecosystem and hugely important for climatic stability. The tropical peat-forests are a major part of this function and they impact through regulatory services on climate, on water management issues, and on soil conservation. They also deliver productive services such as fisheries, game, and timber resources, and they have major cultural importance through indigenous peoples. Furthermore, these biodiverse environments are still not fully documented and unknown sites are still being discovered (Figure 5.3).

6

THE HYDROLOGY AND CHEMISTRY OF PEATLANDS

An introduction to the hydrology and chemistry of mires

The broad issues of peatland water and chemistry are covered in some detail by authors such as Brooks & Stoneman (1997), Wheeler & Shaw (1995), and by contributors to Parkyn et al. (1997). Key early work was by authors such as Ingram (1967, 1982, 1983), and Clymo (1983, 1992). Because of the importance of chemistry and hydrology, it is important to consider how water behaves in peatlands and how this affects the chemistry. Additionally, if we are to understand human impacts on mire systems, these are frequently through actions which modify water and associated chemistry.

The role of water and water-borne nutrients in raised mires is especially intriguing. The only water that reaches the actual central dome of the raised bog is through precipitation, that is, rainfall, snow, fog, dew, etc. Water passes through the mire system, is held within the bog, and leaves by evapotranspiration, surface runoff, or via groundwater. Some rainwater is simply caught by above-ground vegetation and evaporated straight back to the atmosphere. However, most of the rainwater reaches the bog surface and permeates into the peat mass. Some of this water is lost to evaporation, and some is taken up into plants and returned to the atmosphere via transpiration. The combination of evaporation and transpiration is known as 'evapotranspiration' and this process is especially prevalent on hot or windy days.

Water entering the peat system fills the spaces in-between soil particles and other undecomposed organic matter. It is also stored within the large, empty hyaline cells of sphagnum mosses which can hold twenty times their own weight of water. Water held within the soil-pore system is able to move as long as the hydraulic pressure allows it. Hydraulic potential driven by gravitational pull draws water from the high centre of the bog downhill to the outer parts of the peat

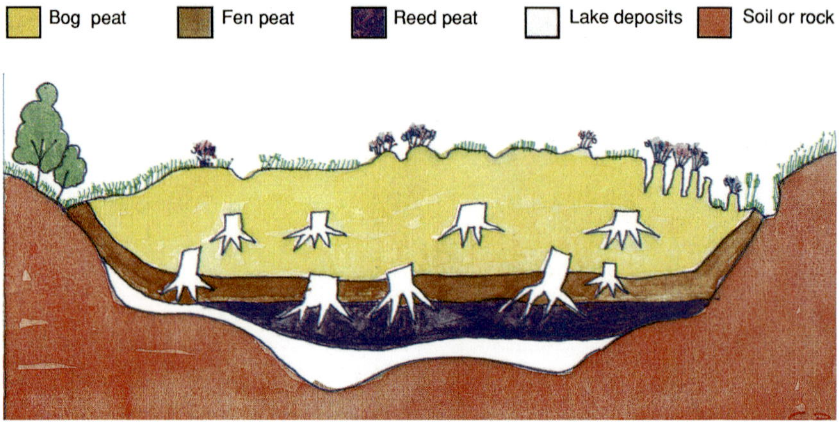

FIGURE 6.1 Section through a raised bog reproduced with permission from John Cross 1989.

dome. The speed at which water moves through the system is in part dependent on its ability to flow within the soil pores. If the soil pores are small or the channels within the peat mass are sparse, then movement may be slow due to low hydraulic conductivity. This water that moves through and out of the bog is termed 'seepage' and represents a loss to the system over time. Seepage can be either vertical and leave via the base of the bog, if the site is not impermeable, or lateral, in which case it exits via the sides and edges of the mire system (Figure 6.1).

Water balance in the bog

For the bog ecosystem to remain hydrologically balanced, the amount of input from rainfall must equal the combined losses from evapotranspiration and seepage, both lateral and vertical. The latter is generally low since bogs mostly form on bases which are impermeable to water. Any changes in the storage capacity of a bog are generally rather slow unless associated with a catastrophic event like drainage or the break-up of a bog surface for some reason. So essentially, the inputs and outputs should balance but with seasonal shifts and occasional variations, especially during wet years or particularly dry ones. Rainfall input should equal lateral seepage and evapotranspiration losses. The surface of a large and relatively pristine bog can sometimes be seen to rise and fall a matter of metres between dry and wet seasons.

$$Pt - Et - Ls - Vs - ChS = 0$$

where Pt = precipitation; Et = evapotranspiration losses; Ls = lateral seepage losses; Vs = vertical seepage losses; ChS = any change in storage capacity of the bog.

As rainfall lands on the raised bog surface, the central dome receives water and this disperses downwards and outwards. This means that the circle of peat around the central core receives both rainfall from above and lateral seepage from the centre dome. By this logic, each concentric ring away from the centre receives more and more water. It is these increasing volumes of water and the associated hydraulic pressure which causes the slope of the bog to increase towards the edge and the whole structure to become dome-shaped. Essentially, the raised mire is maintained by discharge of water which forms a groundwater mound within the 'catotelm' and shapes the peat mass into a hemi-ellipse or dome. A most important consequence of this basic character of the system is that any change to one part of the peat mass affects the hydrology of the entire mire. The peat bog mass forms a single hydrological unit and this is unified and maintained by the lateral seepage of water. There are major implications of this observation in terms of the wider impacts of local human-induced drainage of raised mires; a relatively peripheral drain can impact on the hydrology of the entire system. Historically this impact has probably been responsible for major peat slides or bog bursts associated with destabilisation of large, raised mires.

In the case of European bogs, which are largely formed and maintained by sphagnum mosses, the surface is made up of a tight carpet of mounds of close-packed stems, each growing upwards. The typical hummock-hollow forms carpets of moss and more open, wet pools. Since the sphagnum mosses grow vertically upwards as generally single shoots, they leave behind semi-dead stems and leaves, which become moribund and then die. However, because of the waterlogging and anoxic conditions, the material does not decay but accumulates as peat. This change from living moss to senescence material alters the hydrological conductivity of the system. The main change is that pore-space decreases significantly and so water struggles to move through this layer. Clymo (1992) showed that water movement slows over 600 times in the lower peat compared with that in the upper peat and moss layer (the acrotelm). This means that the peat layer is in effect almost impermeable to free water movement and becomes permanently waterlogged. The effect is that during periods of dry weather, as water levels in the bog fall and drop into the zone of low hydraulic conductivity (the catotelm), further loss of water to seepage is very slow. During wet weather there is a more rapid loss of water, but this is from the upper 'acrotelm' which has more open peat with much higher hydraulic conductivity. These two distinct layers in the peat bog, the acrotelm and the catotelm, are very distinct and have radically different behaviours in terms of water conductivity.

Proctor (1995a, b) provided an overview of the bog environment, observing that maximum summer water deficit is probably a critical factor determining the geographical limits of bog growth, at least at low latitudes. Wheeler & Shaw (1995) further concluded that interactions between precipitation, insolation, temperature, evapotranspiration, and vegetation type are insufficiently understood to provide robust understanding of bog growth and any terrestrialisation sequences. However, it is reasonably well-established that there are close links between regional climate and the overall shape of such domed bogs. In this context, Ingram (1982), proposed the 'Ground Water Mound Theory'

(Ingram, 1982; Ingram & Bragg, 1984). This develops earlier ideas by authors such as Granlund (1932) and Wickman (1951) relating the curved surface of raised bog domes, with structure, function, and vegetation.

The rainwater dome held in the peat mass keeps conditions anaerobic and waterlogged. Such waterlogged, anaerobic peat deposits may be up to ten metres deep from the dome crown, and it is this 'catotelm' which forms most of the peat mass. The deep peat is covered by a thin living layer, or 'acrotelm, around thirty centimetres thick and here the bog water table fluctuates over time. This surface layer is often undulating and is generally called 'hummock-hollow' (Godwin & Conway, 1939; Tansley, 1939), though this is not a strictly correct description.

If conditions are dry, then the acrotelm loses water and water level may drop to the interface with the catotelm but won't fall beyond that. Because of this, as long as the acrotelm surface is within about thirty centimetres of the waterlogged catotelm, vegetation can survive extended droughts. Furthermore, the acrotelm also protects the catotelm from short-term fluctuations in rainfall, sunshine, and from the drying effects of winds (Ingram & Bragg, 1984).

Around the marginal areas of the dome, there are more steeply sloping zones called the 'rand'. This has a more restricted range of peat-forming plants than the flatter central area. As long as the rand is reasonably intact, it will retain some active and protective acrotelm associated with the hemi-elliptical groundwater mound (Ingram, 1982). The latter has a steep curvature towards the dome margins, and this means that the saturated catotelm is close to the bog surface even where it slopes most steeply.

Chemistry in the bog

The chemical properties and characteristics of mire systems are influenced by groundwater and other water inputs including rainfall. These determine the vegetation of the mire ecosystem. In turn, the vegetation of the system together with its microbial flora as affected by water levels and the balance of aerobic and anaerobic conditions dictates the chemical nature of the peat. These conditions are hugely affected and influenced by human activities, and they in turn affect human use of the landscape. The nature and chemistry of mires have impacted on human activities such as agriculture and forestry and continue to have major effects on, for example, water resources and their quality. In Britain, a considerable proportion of our drinking water comes from peat-rich catchments; in Scotland for example, the figure is around 95%.

The chemistry and biochemistry of peat bogs are complex and involve interactions between inorganic nutrients and the organic breakdown products of humification processes. At the core of these interactions is the electrically charged nature of both organic and inorganic molecules – with positively charged cations and negatively charged anions attracted to one another and forming 'bonds'. This means that the abundant organic acids in the peat easily neutralise any incoming alkaline anions such as bicarbonates which are effectively the strongest naturally occurring bases in surface waters. These bicarbonate anions mix in peat and peat-waters with cations of mainly hydrogen and sodium, plus some magnesium, potassium, and calcium. The main inorganic anion in the system is sulphate from the oxidation of organic matter such as amino acids from proteins. However, since the Industrial Revolution, the main sources of sulphate ions have been from acid rain pollution.

In peat bogs it is the negatively charged organic compounds that dominate mire chemistry; organic acids giving the peat-soil and the soil-water a pH of anything from 4.2 to 4.4 in ombrotrophic (acid) mires. Obviously, if there is a strong oceanic influence such as along the western seaboard of the British Isles, then sodium and chlorine are increased through windblown saltwater and marine salts in rainfall. The high levels of dissolved organic matter are evident from the colour of drainage water even from undisturbed bogs, and the naming of rivers and streams such as 'Black Brook', 'Black Water', etc. are a historical testimony to this (Rotherham, 2009, 2013a, b, c). This is different from the effects of major disruption such as agriculture or drainage which lead to a massive release of suspended sediment made up of particles of peat eroded from the mire. It is this latter catastrophic damage which causes the very discoloured peat-waters from damaged mire systems as the organic matter simply washes away.

The chemistry of mire systems is largely determined by humic materials from the peat itself and the water held within the bog. Fens and bogs differ substantially in terms of the base-richness of many fens and the more extensive areas of open water, for example, carr vegetation. In the complex of partial decomposition of the accumulated organic materials there are rich mixes of chemicals from plant matter (mostly) and those from microbial action on the same. This results in macro-molecules from polysaccharides, lignins, and polypeptides which are some of the basic building blocks of nature. Of these chemicals, some are water-soluble and others, such as the polycarboxylic acids (i.e. fulvic and humic acids), are insoluble. These acidic organic compounds may react with things like calcium in animal bones to destroy the skeletal materials from animal remains in the bog. They also interact with substances such as skin, and hair or fur, to trigger tanning processes which in effect help preserve bog bodies of people and animals. One result of these complicated processes is the production of waxy residual materials and these can amount to around 40% of the peat dry weight. This is why, particularly during the nineteenth century, there were many attempts to distil oils or 'paraffin' from peat and also why it can be difficult to re-wet peat once it has dried out.

If a peat bog system is disturbed and begins to dry out then some of these complex macro-molecules become exposed to the air. Oxidation then occurs and they release sulphate, carbon, and ammonium nitrogen into the drainage water. The carbon involved here is mostly colloidal humic material formed from partial decomposition of plant matter. The downstream impacts of these discharges on stream fauna and flora can be severe. This is especially so if the peat-rich waters mix with runoff from farming landscapes which is rich in calcium or iron (from water off limed farmland for example). The effects can look similar to iron-ochre discharges carpeting stream beds where iron oxide contaminates groundwaters.

Finally, one of the other complications is the soil structure associated with peatlands having peat of different depths and with fluctuating water tables. Oscillation of aerobic and anaerobic conditions can lead to mineral iron as oxide coming into and out of solution. Combined with the solution of organic materials leached from vegetation, this can lead to the deposition over time of a hard mineral and organic layer called an 'iron-pan'. This is impermeable and can then trigger future wetting of a site and the development of deeper peats.

Overall, knowledge of water and chemistry in a mire system is central to understanding the ecology of bogs and fens and the physical functioning of the peatland.

PART 3

History, heritage, and people

7

THE PEOPLE OF THE BOGS AND FENS – HISTORY, HERITAGE, AND PEATLANDS

Peat and people

Peatlands trigger radically different responses in different people and communities (Rotherham, 2013b, & c) (Figure 7.1). A Dutch or German farmer perhaps might view a former peat bog landscape as essentially business opportunity and a highly productive agricultural system. Similarly, a farmer in Indonesia or Malaysia may see tropical peat forest as a potential goldmine of opportunity for growing palm-oil crops. For countless generations of people around the world, the peatlands have been a backdrop to their existence and in some cases essential for their very survival.

Even today, peatlands have a far greater impact on our existence and our daily lives than most people realise. Since many peatlands develop at high altitudes and therefore occur around the headwaters of water catchments, they impact on water-flow downstream. This means that peatlands often regulate water flow to rivers, reservoirs, lakes, and farmland at lower levels. They also help control excess water runoff which may cause flooding of farmland, towns, and cities, and during

FIGURE 7.1 Bog body of an executed 14-year-old girl from Northern Germany, 2,000 years old.

drought periods, help maintain the same rivers. Similarly, low-lying fenlands such as formerly in Cambridgeshire, or today in Polish Biebrza, help regulate water-flow to protect farmland from flooding and to filter and cleanse the waters. Both bogs and fens provide habitat for often rare and important wildlife species.

Peatlands as richly cultural landscapes

Globally, peatlands are often closely associated with people and have been so since prehistoric times. Whilst ceremonial or sacrificial burials in bogs and the resulting 'bog bodies' have captured imaginations, the wider cultural heritage and histories of peatlands are less recognised. Yet unless the wetlands have been totally eradicated (which sadly is quite often the case), the human footprint, and the cultural heritage are imprinted deeply into the landscape and the communities.

Humans have been active in and around bogs and fens for millennia, and the archaeology of the remaining sites is incredibly rich. In Europe, for example, the peatland cultural landscape extends back to times when peat was first forming through Romano-British periods, the 'Dark Ages', into the medieval, and so through the early industrial to the modern-day. In all these major periods of history and prehistory, people have lived in and around bogs and fens and they have left their marks as evidence. Yet despite this, many people interested in peat bogs and fens are relatively ignorant when it comes to their cultural history. The heritage and archaeology along with the palaeo-archaeological resources are mostly overlooked or at best, poorly understood even by many people involved in peatland restoration (Figure 7.2).

With the results of many archaeological investigations and widespread palaeo-ecological and palaeo-botanical studies, it is possible to 'reconstruct' models of past landscapes and their people. Through such investigations we can gain good insights into the human inhabitants of peat landscapes over the millennia and importantly, their impacts on the environment through long periods. Significantly too, the bogs and fens hold evidence of past climates, biodiversities, and of changes over time. If we want to understand the processes and implications of climate change, then looking into the sediments of mire systems is a good starting point.

In times past, settlements around fens and bogs moved as the waters rose and fell. For some, these landscapes or perhaps 'waterscapes' provided rich supplies of food, fuels, and building materials, and sanctuary from persecution or conflict. For others, such places were watery and disease-ridden 'wastes' to be avoided or better still, to be tamed. In Great Britain, for example, peatlands communities were often independently minded and frequently were outside of mainstream society. Wetlands were places of sanctuary for outlaws and others at odds with the church or the state, and they also gave birth to non-conformism. Methodism, Quakerism, and the English Civil War are just a few examples of social movements and events which sprang from the English fenlands (Rotherham, 2010, 2013a, b).

For early communities with relatively few possessions, the rise and fall of water levels over the centuries was less problematic than it is for later, more

FIGURE 7.2 Peat bogs and fens are rich in artefacts such as this prehistoric dugout canoe from Halmstad Sweden.

settled and sophisticated people. Settlements could simply move up or down the hill as conditions allowed. However, once the countryside was more permanently settled and more densely and territorially inhabited, then fluctuating levels of floodwaters became rather more problematic. Human society tends to expect nature to be stable and constant, which of course it is not.

War and peatlands

Across the world, fens, bogs, and other wetlands were also places which influenced conflicts and politics, through independent thinking as noted above, and through strategic and topographic situation. Flanders Moss in central Scotland for example was hugely important in the control of the political and military connection between the Highlands and the Lowlands of the country. This positional and political importance has meant that the ownership and control of peatlands have been important throughout history. Most recently, for example, the drainage of their wetlands was a key ploy in the war of Saddam Hussein against the Marsh Arabs. Similarly, as described by Grunwald (2006), the Native Americans held out in the Florida Everglades when under a final assault from the European North Americans during the later 1800s.

Landscapes transformed

Communities living in and around peatlands altered them through drainage, control, and management of wetlands shaped and manipulated to human functionality. People adapted to life in the wetlands, and the peatlands were

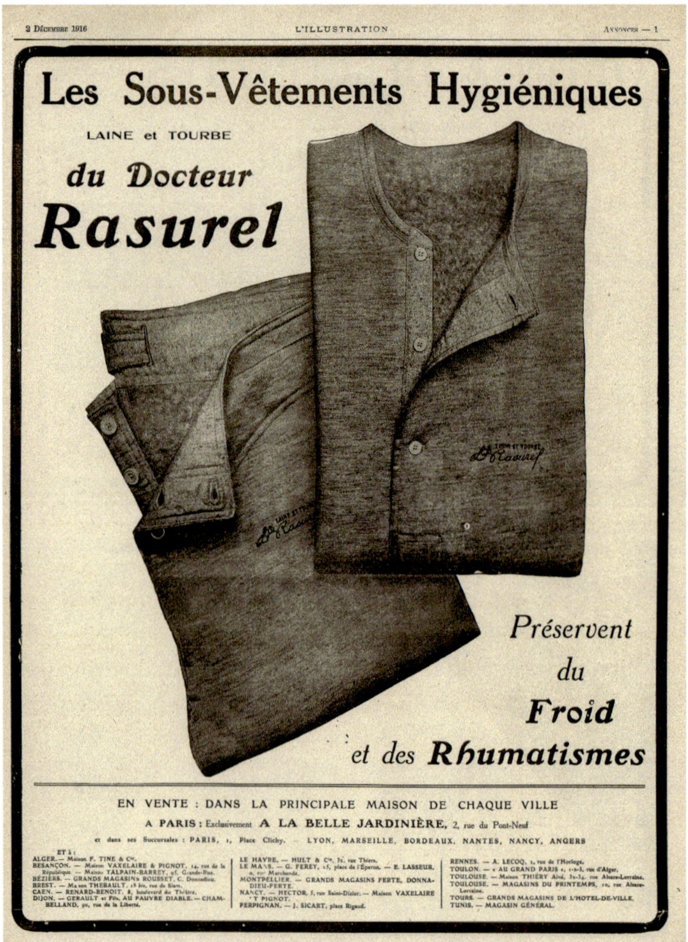

FIGURE 7.3 Peatlands provide valuable resources for local communities including material for clothing such as thermal underwear from peat fibres as in this case from France in 1916.

modified by human usage and activities. Generally, however, as human impacts grew, the peatlands shrank and dried. Whilst there is an extensive literature of local histories which refers to peatlands and their exploitation and uses, there have been few accounts specifically about this fascinating aspect of the 'peatscapes' past and the people that lived there and interacted with wet nature (Figure 7.3).

8

PEATLANDS AND HUMAN CONFLICTS

Introduction

Historically peatlands have been of importance and significance in the landscape and to people worldwide. The scale of the importance of these landscapes has generated remarkable relationships throughout history including through human conflicts and stresses, particularly in times of war (Rotherham, 2009, 2010, and 2013a, b). This has impacted on peatlands, and they have influenced the outcomes of conflicts. In Roman times, the fens and bogs were brought under varying degrees of control through ambitious drainage schemes, but the primary function of these waterways was probably military – which was moving men, animals, and supplies speedily through otherwise difficult terrain. Manipulation and drainage of peat landscapes for military transport is one example of passive usage. Some of the issues and observations are discussed in my three books, *Peat & Peat Cutting* (Rotherham, 2009), *Yorkshire's Forgotten Fenlands* (Rotherham, 2010), and the *Lost Fens* (Rotherham, 2013b), and were expanded in the 2013a volume '*War and Peat*'.

Examples from northern England

From the 'battles' for the Peak District's Kinder Scout and the recreational access to the moorlands to saving South Yorkshire's Thorne Moors and Hatfield Chase, peat bogs and moors have been evocative places for local people. These are truly contested landscapes. Not only this, but they played major roles in the war effort of the two World Wars. Forgotten wartime aspects of peat bog and fen include uses such as the supply of vital horse-litter (cut peat-turf) from sites like Thorne Moors in South Yorkshire.

In the 1600s, beginning in the Northern English Fens but spreading southwards to Cambridgeshire, conflict over the proposed drainage of the Fenlands was a key factor in triggering the English Civil War. Clearly, people and peatlands interact in many complex and complicated ways. Furthermore, human conflicts affect and are influenced by these once vast landscapes. I suggest there are perhaps seven main ways in which people and peatlands have interacted in times of conflict or other stresses. Examples of these given below and some of the uses and interactions are then developed further.

Impacts, utilisation, and examples

1. Peatlands influence war and conflict through topography and nature, often being of critical significance to strategy and outcome

   Examples: The military significance and use of moors, heaths, fens, and bogs extend over millennia. For instance, the moorlands and bogs around the western edge of Sheffield can trace a military presence across millennia from Iron Age hillforts to the twentieth century. In modern times, they were used both as military training grounds and during World War Two as decoy areas ('starfish' sites) to protect the vital steel industry in east Sheffield from German night-bombers. Moors and lowland heaths also provided many sites for the 1940s airfields for the 'Battle of Britain' and then for later air strikes against Nazi Germany.

   However, there is more, since wetlands, especially peat bogs, fens, and moors, were hugely significant in many military campaigns. In the military landscape, these treacherous areas of neither firm ground nor navigable water were a serious problem for soldiers or warriors, most of who could not swim anyway. In early conflicts, a slip into a fen or a bog wearing armour or chain mail, for example, caused one to sink slowly to a very unpleasant death. The Saxon troops of Hereward the Wake, holding out on Cambridgeshire's Isle of Ely, drove the Norman invaders off the fragile fenland causeways by firing the reeds. The Norman soldiers in their chain mail sank deep into the quagmire. At East Yorkshire's Stamford Bridge, an earlier Saxon-Viking battle was settled in a vast wetland landscape in which the one river-crossing was a vital and strategic key. Later in medieval times, this aspect of wetlands in the landscape was used to great effect by the Scots against the English at the battles of Stirling Bridge and Bannockburn. Similarly, battles at Solway Moss, Culloden, and Flodden all turned on the impact of a bog or wet moor.

   The massive raised bog of Flanders Moss, together with the meandering River Forth, was of huge military and political strategic significance in Scotland. It was said that whoever controlled Stirling Castle, the gateway to the Highlands, controlled Scotland. In England, many of battles of the civil wars, such as the Wars of the Roses, and the English Civil War, took place on some heath, moor, or bog. The Battle of Sedgemoor itself was a tragedy played out in a foggy, dank marshland close to where the Saxon King Alfred the Great had once sought refuge from Viking incursions deep in the wetlands of Athelney.

In the European arena of the two World Wars, peat bogs and fens played major strategic roles. In World War Two, as they retreated, the Dutch resistance, for example, flooded the former peat-cuttings to halt the German advance. The Germans later retaliated by flooding the agricultural polders (reclaimed from the peatlands) to destroy the Dutch crops and cause mass starvation. The First World War battle of the Somme and other similarly catastrophic campaigns were fought in the vast wetlands and peat bogs of the Low Countries.

2. Heaths and moors were often used as military training grounds, from early times up to the present day

   Examples: Around Sherwood Forest in England, the famous heaths of the old royal forest became military training grounds, and in World War Two, Clumber Park in the Forest was a transit place for tanks and was used as a camouflaged storage site. Dartmoor, the Peak District, Cannock Chase, and other British heath or bog sites were used for military manoeuvres probably as far back as the Crimean War.

3. Peatlands were used as wartime bases especially during World War Two when they were brought into usage as airfields and similar facilities

   Examples: In Nottinghamshire, Lincolnshire, Norfolk, Suffolk, and East Yorkshire, commons, heaths, and fens were turned to use as airfields – some still surviving to date. Moorlands and heaths provided the sites for airfields for the 'Battle of Britain' defence and the offensive strikes against Nazi Germany.

4. Bogs, moors, and fens especially, had unplanned and passive roles in warfare, such as being the last resting places for stricken or lost bombers of both sides during World War Two

   Examples: These landscapes became the last resting places of aircraft and crews that crashed on Kinder Scout and Bleaklow in the Peak District uplands, and on South Yorkshire's Thorne Moors in the lowlands. This is a tragic history, which should not be forgotten. According to German colleagues, peat bogs were sought as places of relative safety where families could take refuge when Allied bombers returning from night-time missions jettisoned unused bombs. The peat absorbed the explosive impacts better than dry land. Similarly, Italian soldiers in World War One found peat bogs to be good places in which to shelter from shelling.

5. Peatlands, especially bogs and fens, provided materials of immense value to wartime efforts such as supplying horse-litter, sphagnum moss, alder buckthorn, or peat fuel

   Examples: There are several examples, now mostly forgotten, of peatlands supplying vital resources for military use or else for wider society during times of conflict. For example, early twentieth-century wartime resources taken from peat bogs and fens included the supply of vital horse-litter (cut peat-turf) from sites like Thorne Moors in South Yorkshire, or the harvesting of sphagnum for wound dressings. The latter occurred widely across Britain and Europe and in North America, during both World Wars.

One of the main uses of peat in the late 1800s and early 1900s was as litter for animal bedding. With huge numbers of animals powering pre-petrochemical farms, towns, and cities, there was a big demand for material to keep things clean. Peat was ideal for this purpose, and once soiled, its nutritive qualities were enhanced and could be used on the land as vital fertiliser. At Thorne Moors in South Yorkshire, the English Moss Litter company extracted peat moss up to the 1960s; and from 1923 to 1962, the Midland Litter Company took moss from Fenn's Moss near Wrexham. Although raw peat was widely used as litter by farmers and peasants the wider usage developed more significantly in the late 1800s and early 1900s. This was firstly for supplying litter for horse-drawn transport in towns and cities and then in the World War One war effort. Processed and packaged as a commercial product, during the First World War, with its absorbent and fibrous nature combined with antiseptic properties, it was used extensively as military horse-bedding. Peat was also used as animal feed, mixed with green fodder and perhaps molasses. Again, its antibacterial properties may have had therapeutic effects and it was used either coarse or as a powder mixed into cattle cake.

To give some indication of the emerging demands for peat-moss litter, there was an increase of around 98,353 working horses in Britain between 1901 and 1906. These were employed by railways, tramways, omnibus companies, various local authority undertakings and of course many other businesses. Peat-moss litter (dried peat) made an ideal bedding material and by the late 1890s, at Thorne Moors, the British Moss Litter Company was formed and took over various established peat works across the area between the Rivers Don and Trent. Alongside the system of canals and boats for carrying peat, there developed a network of narrow-gauge railways and external connections to the wider rail network. Established as a major centre for moss-litter production and employing at its peak around 350 men, South Yorkshire's Thorne Moors produced vast quantities of peat-litter for the many horses sent to the World War One front line and which had to be catered for. Peat-litter is the untold part of the 'war horse' story.

The demand was for litter and a diversity of other products with a new 'pressing mill' was being built in 1895 and is still known as the 'Paraffin Mill' or the 'Paraffin Works'. This was to produce gas for fuel, ammonia water, paraffin, creosote, methyl alcohol, tar, and alcohol for motor cars. Peat dust was used to pack fruit and peat was even fed to cattle. However, these diversifications did not last long beyond the end of World War One and the mill closed in 1922. With the declining use of horses for industry and transport, the moss-litter business also collapsed.

Peat as fuel was the earliest use and horticultural peat use was an afterthought. The latter initiative brought catastrophe to the South Yorkshire peat moors and bogs as it did elsewhere. Here it triggered further conflicts over resource use with battles by one of the world's first direct-action environmental groups, 'Bunting's Beavers', and by others. These were the attempts to save the remnants of the once vast South Yorkshire lowland moors and ultimately led to the National Nature Reserve present in the area today.

The region's moors and bogs were also used to harvest sphagnum moss for wartime medical and surgical uses. During World War One, as discussed later, this hugely important activity saved countless lives of injured servicemen.

6. Conflicts, socio-economic stresses, or extreme weather all influenced fuel use and competition and therefore peat exploitation

   Examples: The use of different fuels for energy-supply varies with availability, accessibility, suitability, and cost. In many areas, throughout human history, peat was a very important fuel. However, even when peat-turf had been displaced by other more convenient energy sources, it might come back into use during times of stress and other crises such as conflicts or post-conflict shortages. In this context, often forgotten are the effects of post-conflict scenarios, such as in Britain after World War Two the resulting energy crisis for domestic and industrial fuels – for example, at Holme Moss, in the South Pennines, there is one of England's last community turbaries (a legal peat-cut for domestic fuel). Here, two former soldiers arrived after World War Two to set up a business. Recently discharged from the army, they used their discharge money to set up a peat fuel business and for several years in the post-war energy crisis, they supplied peat fuel to factories as far away as Sheffield and Leeds. Similar exploitation affected the post-World War Two blanket peats of the south-west Pennines as large areas were cut for peat fuel to supply Pilkington Glass in Lancashire.

   War could be devastating for peatlands for several reasons. It often led to shortages of food as well as fuel, and in Britain, the World Wars seriously affected overseas imports and triggered the need to be more self-reliant. This meant, on the one hand, intensification of land-use, for example, World War Two brought about the final demise of the remaining Southern Fens of eastern England. On the other hand, the conflict could lead to the abandonment of for example drainage schemes. At Walberswick in Suffolk, marshland drainage was unmaintained and this allowed reversion to fen and marsh. Around Leighton Moss in Lancashire, the long-standing peat-cuttings were abandoned as the site re-wetted due to neglected drainage. The ultimate result in both cases has been the development of major nature conservation sites by the twenty-first century. These are to a very large degree the inadvertent by-products of the wars.

7. Fens and bogs as places of sanctuary, of non-conformism, and independence

   Throughout history, wetlands such as peat bog and fen have provided sanctuary for people in times of conflict or oppression, and for non-conformists and others seeking to distance themselves from the law or the church. Rich in natural resources for those who knew and understood their ways, but difficult to enter or transverse if you did not, these were ideal hideaways from the 'Dark Ages' times of Alfred the Great, to Hereward the Wake during the Norman Conquest of England.

Other such examples are the nineteenth-century French forces in the Franco-German war, to the Marsh Arabs of Iraq seeking protection from Saddam Hussein. One consequence of the sanctuary that wetlands gave from those in authority was that those in power sought to remove them from the landscape to control both the environment and the people. The English fenlands were communities of independent thought and action. During the later Middle Ages and Early Modern Period, they gave rise to such radical movements as Quakerism and Methodism. Alongside a relatively wealthy and independent peasantry in the English Fens, there was competition for resources between the commoners, the local gentry, the wealthier aristocracy, the Church, and the Crown. In England, one of the outcomes of competition between these vested interests was the English Civil War in the 1600s, a cataclysmic event which changed global human history (Figure 8.1).

FIGURE 8.1 The famous *turfschip* of Breda in 1590 was the Dutch equivalent of the Trojan Horse used by the ancient Greeks. In this case, the ruse was employed to breach the fortified town and free it from the Spanish overlords.

FIGURE 9.2 Sphagnum mosses in the Burbage Valley of the English Peak District and a rich target for the wartime moss collectors.

Conflict and energy supplies

Sometimes, war or extreme weather might trigger a move from the utilisation of say coal, back to peat, a resource either not used or used historically but then abandoned. This was described for farmers in the south Lancashire Pennines and also in Cumbria, apparently as a response to bad winters and coal not being available. They reverted to using the local peat turbaries. This seems problematic since to be useful, the peat-fuel needs careful harvesting and drying. In a bad winter, peat would not solve the immediate problem since you have to plan six to nine months ahead. It seems more reasonable that the farmers reverted to peat use during the wartime period or the 1920s 'Depression' as a response to economic pressures and energy shortages.

However, this phenomenon occurred around the world, especially, for example, in North America. This was mainly along the eastern seaboard of the Atlantic but elsewhere too. A considerable literature from the 1800s into the 1900s addresses the potential of the USA and Canadian peat-fuel resources. Peckham (1874) for instance, considered the potential of peat and turf for domestic fuel supply in Minnesota. A key point that he made was the importance of competition with, and a scarcity of, more sought-after fuels. Peat-fuel might be viable if an area was remote from forests, and away from coal transportation routes. Shaler (1895) examined the origin, distribution, and commercial value of the USA peat deposits, noting the relationship between peat use and lack of available fuel-wood. In northern Europe, he suggested the greatest use of peat-fuel was during the eighteenth century when forests had been cleared but with transport to rural areas difficult, mineral coal was not in widespread use. Most rural communities in Northern Germany, Scandinavia, Russia, France, and the British Isles, depended on peat and turf for household fuel. The only exceptions would be in the case of wealthier classes who could afford to buy other fuels and had the technology to burn them. Shaler described the descendants of Native Americans in the town of Gay Head being some of the last to use peat-fuel in that region.

In North America, by the late 1800s, availability of cheap anthracite coal led the move away from peat. The decline in peat use (free except for the cost of collecting), was caused by the availability and low price of coal. The first commercial extraction and processing of peat-fuel in the USA was then in 1902 when a strike by Pennsylvania miners caused a fuel crisis (Soper & Osbon, 1922). However, in the USA, for example, during World War One, there was considerable interest in a move from coal and wood to peat or turf as domestic fuel. Turp (1916) reported that very little machine-processed peat-fuel was extracted in the USA in 1915, but by 1917, operations resumed.

Haanel (1918) noted difficulty obtaining adequate and cheap fuel supplies in Canada. The reasons related to war were complex including USA labour shortage that restricted coal imports to Canadian provinces. The situation was considered very grave in that fuel could be reduced or even cut off completely. The report argued for urgent attention to developing peat fuel as "an excellent substitute for coal". He finishes the paper with a plea "....... to the establishment of a peat industry on a sound basis in Canada, and thus insure the people against a possible shortage of fuel and the suffering it would entail". Blizard (1917) reported on the value of peat-fuel for industrial steam generation and considered the key issue to be price competition with coal. As coal prices rose, peat would become competitive. According to Soper & Osbon (1922), the wartime coal shortages in the USA in 1917 and 1918 showed peat-fuel could be extracted and be competitive if coal was scarce and expensive. This observation is supported by subsequent activities and by numerous governmental and scientific reports.

The Irish question

Traditional peat-cutting has been of huge importance to the Irish people and cutting for domestic fuel has caused the greatest decline in Northern Ireland peatlands. With the demise of native woodlands, peat became the major source of fuel in Ireland during the seventeenth and eighteenth centuries. Rights to cut peat on small plots of land, known as turbary rights, were allocated to landowners and traditionally peat was cut by hand using a special turf-spade known as a sleán or slane. Interestingly, hand-cut turf production in Ireland reached its peak in 1926 when over six million tonnes was cut. This was during the Great Depression and followed World War One, and the Irish 'Troubles'. The amount of turf cut then declined steadily until World War Two, when peat became a vital domestic fuel source again as coal supplies from Great Britain almost ceased. Many of the deep-peats in Irish raised bogs and blanket bogs were extensively cut.

In Ireland, the use of fen-peat as a source of fuel, known as mud turf, was less common because the peat is very shallow and cannot be cut with a spade. Instead, mud turf is gathered by digging a hole and mixing water with the peat, then tramping or 'puddling' it with bare feet, shovelling it onto the bank and finally moulding it into blocks by hand. This was a very labour-intensive process and only practised in a few areas, such as Brackagh Moss in County Armagh. After the World War Two, the low price of coal and oil kept peat-cutting to a minimum, and by the 1970s, the annual production of peat was down to about a million tonnes, mostly from the blanket bogs in the west. However, the introduction of tractor-drawn auger machines during the 1980s increased the amount of peat cut again. Since then, mechanised peat extraction has become the norm in Northern Ireland and the tradition of hand-cut turf has almost disappeared.

In Northern Ireland, 78% of raised bogs have been cut for fuel, mostly for domestic purposes. Yet the relatively limited commercial extraction for fuel has had important local effects. Commercial extraction needs planning consent, but the complexity of land ownership and turbary rights sometimes makes it difficult to distinguish between commercial and domestic cutting. This brief description of the situation in Ireland demonstrates the overall impacts of utilisation, the various approaches, and the potential effects of war or other socio-economic stresses. The impacts of commercial peat-cutting by the Irish government energy agency, Bord Na Mona, have been disastrous for the Irish landscape and the associated peatland ecosystem services.

War and the English Fens

With the outbreak of World War Two, there was an upturn in interest in farming improvement and drainage by the British central government. In the Fenlands as I described in '*The Lost Fens*' (Rotherham, 2013b), the land around Wicken Fen in Cambridgeshire was acquired by the pioneering horticulturalist, Alan Bloom. By the spring of 1940, he was considering ambitious plans. "There were other more or less waterlogged and inaccessible parts of the Fens, and why should I not try to get a few people together with money to invest, and make a big thing of this reclamation job"? His scheme did not get much support but it was a hint of things to come, and by June 1940, he had invested in a caterpillar-tractor to access wet, difficult land. Bloom described how his Priory Farm had become a battlefield on which the forces of dereliction had paused in their encroachment on the farmed land.

He felt the advantage had for a long time been with 'wilderness' and nobody could farm against water. On the one side 'swamp' was deliberately encouraged because it gave sanctuary to wildlife, but now with the tables turned, the plough was on the offensive for farming and food production. The level of water in the dykes, which divided the two, determined which would prevail, and with flat land on both sides of the drain, there was no possibility of compromise. Bloom's personal account (Bloom, 1944) gives an insight into processes and drivers in this bitterly contested landscape close by today's Wicken Fen Nature Reserve. He felt it was a "fight the National Trust, or rather, I supposed, the local governing Committee......, men with mainly academic interests. They could not see things in the same light as those whose interests were agricultural". The final stage seemed to be set on potentially bringing into cultivation around fifteen acres of the flooded Adventurers' Fen as requested by the War Agricultural Committee.

However, this situation speedily changed when it was decided that the site was to be used as a bombing target. Both Bloom for Priory Farm, and the National Trust for the nature reserve, received the appropriate requisition notices. But the situation soon changed abruptly once again when the requisitions were cancelled. By August 1941, the Biology War Committee presented a Memorandum to the Joint Committee of government research organisations on "R. frangula as a source of charcoal for munitions". The plant *Rhamnus frangula* or *R. catharticus*, (alder buckthorn), was long-known to make fine grade charcoal for explosives. Unfortunately, the species is uncommon in Britain and with maritime blockades, overseas supplies had stopped and this was an urgent matter of national security. The report triggered a survey of the plant's distribution (November 1941), and it was found on Adventurers' Fen. Reclamation work was temporarily halted with Bloom and other farmers commissioned to clear scrub and selectively harvest the alder buckthorn, and for this the National Trust got paid.

The work involved cutting or coppicing the buckthorn with anticipated regrowth over three years. Within a few months, the buckthorn was all cut out and the War Agricultural Committee was anxious to bring the Fen into cultivation. The proposal was now for all the 286 acres of Adventurers' Fen since ".....the Catchment Board engineers and officials, the overseers and their men – were of the opinion that Adventurers' Fen could be and would be drained". Some local farmers still felt drainage would be a colossal waste of money and may not work. However, the environmental battlefield received a visit from the War Agricultural Executive Committee chairman, and soon after, the King, the Queen, and others.

The scene was set and the War Agricultural Committee had already started reclaiming the southern area of Adventurers' Fen before March 1941. They hoped to be cropping the whole site by summer that year. Catchment Board men had deepened the drain immediately north of the nature reserve, and well into the core-site the lowered water table impacted considerably.

Across the Fen the National Trust had taken action to make the whole site wet. This was ideal for conservation but a major hindrance to the drainage and improvement of Burwell Fen and surrounding areas. Major new drains were cut across the ancient landscape to release the pent-up waters and Bloom wrote, "It was sheer delight to watch that water running full pelt from seven or eight points along the boundary of the Fen out of those grips we had been digging". After a drying breeze over some days, they burnt off surface vegetation as lighting just a small pile of sedge litter resulted in flames that spread as if the area was doused in paraffin. Within less than thirty minutes the Fen completely changed its appearance, the dull buff-grey turned to charred black with burnt bushes and smouldering sedge hassocks was all that remained.

The cultivation costs were grant-aided at £2 an acre plus the cost of ploughing. Half of the grant covered costs and the other half went to the landowner or farmer with the tenancy running for three years after the end of the war. As Bloom admitted, by late 1942 the costs of the work were clearly underestimated, but "……there was nothing I could do about it. We had our work cut out to get the two hundred and eighty-six acres cropped by 1943, let alone 1942". King George and Queen Elisabeth, the Minister of Agriculture, the Duke of Norfolk, and Mr Tom Williams MP, soon visited the site with the Press to inspect the efforts to feed the besieged country.

But by February, the land was losing much of its topsoil and losses were progressively worse through March. Even so, the area was soon sown with oats, barley, beet, and potatoes. Then, with dry weather came the ironic spectre of drought, and the problem being alleviated by abstracting from Wicken Lode caused further depression of the water table on the remaining nature reserve. During May, a gale tore up great black clouds of dust over Swaffham, Waterbeach, Soham, and Isleham, and the sky turned "a queer, dark tinge for hours" – the worst 'blow' for years and a direct result of the War Agricultural Committee's improvements. The dust settled over wide areas up to twenty miles away; the desiccated ancient peatland was just blowing away. The wind-blown peat even infilled the drain which needed to be re-dug and some crops needed replanting two to three times. By summer the farmers cropped wheat, barley, and sugar beet.

By late 1943, with initial work done, the land moved towards intensive, industrial agriculture. Bloom noted that

> Adventurers' Fen and Priory Farm had proved that crops equal to any other black fens – and better than some –could be grown. Those ideas and hopes that for so long I'd been pushing back into the pigeon-holes of my mind, could now begin to emerge. More complete fertility, extended mechanization, more and better buildings, a thorough livestock policy, alternative leys to give some of the much-cropped land a rest in turn.

Improvement of these 300 acres had cost a lot but Bloom's view was that it was the country's fault for neglecting the land in the first place. The intensive two years simply made up for the inaction of twenty or thirty years before and he now looked forward to the permanent recovery of British agriculture and an end to neglect. He felt "……the welfare of the land must in future run parallel with that of the nation", and people demanding that, "…..cheap food, abundant in quantity and variety, is the only thing that matters". This vision oversaw the final demise of ancient fenland in southern England in tandem with rapid mechanisation and agri-industrial development fuelled by post-war zeal for food self-sufficiency. The public purse and petrochemicals transformed the countryside and its communities. The peatlands were gone or in terminal decline.

The last of the old fen

In 1954, a contemporary of Bloom, the popular author, broadcaster, and fenman, James Wentworth-Day wrote his '*History of the Fens*' (1954) that Bloom had come to 'improve'. He described how

> .... in the witch-hours before dawn, the smell of the fen. A strange indefinable smell, scent of reeds and peaty waters, of sallows, and meadowsweet, of rotten lily pads – and of fish. That smell of freshwater fish which is penetrating, ineluctable, indefinable. An old, strange, blended smell, a smell as old as Time, compounded of scents that belonged to an untamed, undrained England, the England of the Saxons.

Wentworth-Day wrote of the whimper of wild ducks' wings at night, with the thin whistle of the teal, and the pig-like squeal of water rails. There was the *kerk-keek* and *ker-erkk* of moorhens moving from lode to lode at night, and the *br-ooomp-oomp*, hollow and ghostly of the bittern in May and June, the quiet chorus of secret fen voices during the 'manless' hours carried on the soft fenland breeze.

By the end of World War Two this evocative peat-based landscape was reduced to the rump of Wicken Sedge Fen, all else was gone. Nevertheless, though Wicken remained undrained, it was not unchanged, but was

> Still a place of dense reed-beds, of sedge jungles, of forests of sallow bushes and creamy oceans of meadowsweet. But the old village proprietors, the fen owners, who each had their few acres of the wild fen, where they cut their reeds, mowed their sedge, and speared their eels, have sold out.

Wentworth-Day noted how the National Trust, now as owners of the Fen, cut neat grass rides and placed signs on neat white posts to tell you where to go. The villages had changed with mud and thatch cottages tumbled down back into the earth from which they came, replaced by "hideous villas of staring white Cambridge brick, with their grim, unsmiling roofs of alien slate, under which no swallows nest, on whose rooftrees no starlings whistle". The villages around Wicken and Burwell were "......divorced from their brown and smiling mistress, the fen. And the villages are the poorer".

In 1935, Wentworth–Day had bought part of the old Fen "....a half-drowned, stinking swamp of disused peat diggings, reed-beds, and interlacing dykes". He stopped up the drains to hold back the winter floodwaters and meres were instantaneously recreated. This miniature oasis close to the remnant Wicken Fen quickly drew huge numbers of wildfowl with a diversity of water-birds both common and rare. Wentworth-Day recorded pintails (*Anas acuta*) and golden-eye (*Bucephala clangula*), common terns (*Sterna hirundo*) and arctic terns (*Sterna paradisaea*), six cattle egrets (*Bubulcus ibis*) and a great white egret (*Ardea alba*), mallard (*Anas platyrhynchos*), teal (*Anas crecca*), garganey (*Anas querquedula*), gadwall (*Mareca strepera*), shoveller (*Anas clypeata*), curlew (*Numenius arquata*), curlew sandpiper (*Calidris ferruginea*), green sandpiper (*Tringa ochropus*), common sandpiper (*Actitis hypoleucos*), greenshank (*Tringa nebularia*), a yellowlegs (*Tringa* sp), bar-tailed godwits (*Limosa lapponica*), ruffs and reeves (*Philomachus pugnax*), little grebes (*Tachybaptus ruficollis*), common snipe (*Gallinago gallinago*) and great snipe (*Gallinago media*), and much more. He had starlings (*Sturnus vulgaris*) coming to roost in flocks half a mile long and a 100 yards deep; there were even nesting black-necked grebes (*Podiceps nigricollis*). Hen harriers (*Circus cyaneus*) and marsh harriers (*Circus aeruginosus*) swept over the reed-beds and Montagu's harriers (*Circus pygargus*) bred there. This gives just a flavour of a lowland peat site in full glory.

"A wild and lovely place, which dwells in the memory as a very perfect picture of the older England, the England of Hereward the Wake and St Guthlac, the Saxon hermit." On Wentworth-Day's little fen "Coots clanked, ducks splattered, snipe drummed, pewits wailed, and the redshanks sprang on flickering wings, ringing their carillon of a thousand bells", and up to 50,000 sand martins (*Riparia riparia*) swirled in massive migration roosting flocks. It was, in just this few 100 acres of Adventurers' Fen, "....the old spirit of the Great Fen that once covered half Lincolnshire and Cambridgeshire", rose but destined not to last long.

Then with World War Two,

> They drained the fen with a great clamour of bureaucratic self-praise. The waters went away and the fish died by the cartload. The reeds stood rustling and dry above the black mud. Then they set fire to the reeds, and for a day or more my secret fen roared and crackled in a tawny yellow, red-hot sea of flame. Great billowing clouds of black smoke rose up and polluted the blue skies and swept away on the wind until dust, ashes, and smoke fell like a grey pall on the roofs and the green heath of Newmarket, away on its windy upland.

The duck rose up and were away, and the moorhens, rails, bitterns, warblers, and others too. When the wind blew away the dust, smoke, and ashes, all that remained of the secret fen was

> ....burnt and black and scorched. An insult to the high fen skies. An altar of burned beauty. A sacrifice to man's neglect of pre-war farming, a burnt offering to humanity's failure to live together in harmony. And thus, in a funeral pyre vanished the last and loveliest remnant of what had been a recreation in all its wild glory of the ancient Fens of Eastern England.

Wentworth-Day had bought the fen

> ... to preserve it, to save for all time the essential Englishness of it, to love and enjoy the sight of birds and clouds, the wind in the reeds, herons fishing in summer shallows, gulls wheeling against May skies, the sting of winter sleet ......

Now it was no more and he questioned the wisdom of it all.

> Is the world any better for this change in my fen, or in the ten thousand acres of other fens which they have drained, burned, grubbed up, and cultivated during the War? Materially, yes. Spiritually, no. Economically, again no. Those are the answers in a nutshell. On my fen they spent thousands of pounds in expensive drainage, in constructing concrete roads which will probably crack, sink, and become derelict in a few years. The bill for our County Agricultural Executive Committees is estimated to be in the neighbourhood of £25,000,000 a year. Do the Committees grow £25,000,000 worth of food each year? The answer, I think, is no.

The war brought a brief reprieve from total destruction to allow alder buckthorn harvesting but ultimately the fate of the Fens was sealed. In the wider landscape, arable production from old peat-fens rivals the best in the world; though it depends on massive inputs of petrochemical fertilizers and fuels. It is arguable as to whether the wartime 'improvement' of the relict fens produced much at all in proportion to the substantial investment. Today however, we know that the productive lifespan of these 'bread-basket' landscapes is pitifully short and there are fewer than fifty harvests left in the soil (e.g. van der Zee, 2017). Yet this little story in the Cambridgeshire Fens has been repeated across peatland landscapes all over the world as ideas and technologies were exported from Britain and Europe (Rotherham & McCallam, 2008), that is still happening. Furthermore, in the intimate personal accounts of Bloom and Wentworth-Day we capture the essence of the fate of peatlands globally.

The particular example of fuel allotments and common rights

Throughout history where peatlands occur, local communities have exploited them for peat-fuel. When land improvement has taken away commoners' rights to take fuel from the bog, then in some cases at least, special 'fuel allotments' were set aside either or both as compensation for rights lost or for the subsistence of the landless poor. The story of the parliamentary enclosures and improvements to English commons captures an element of this wider transformation.

Frimley Fuel Allotments, Surrey

A good example of a peatland fuel allotment and one particularly affected by military use is provided by the Frimley Fuel Allotments Charity, founded in 1801. This is particularly well documented and provides unique insights into provisions for the upkeep of the poor during the English enclosures. The Fuel Allotment Charity owns the land on which the Pine Ridge Golf Centre is built, along with around 100 acres of open access heath. The Charity was established when the Parliament under King George III passed the Frimley Enclosure Act in 1801. The common was physically enclosed in 1826 with a portion set aside to provide 'Fuel for Firing' of the poor of the Hamlet of Frimley. This is documented by Wellard (1995). A perhaps unique aspect is the wealth generated by the arrangement for the golf course on part of the land-holding, providing an income of £65,000 per year by the 1990s.

In 1793, the extensive 'wastes' of Frimley supported only a small population of local people. From that time to date, almost 90% of the modern settlements of Camberley, Frimley, and Frimley Green were built on what was then an expansive tract of open heath and bog, including Frimley Heath, Cow Moor, Bisley Common, Pirbright Common, and Chobham Common. The area was covered by gorse or furze, heather, scrub, and sparse grass with abundant deer and poor grazing for sheep. Along with access to turf fuel, commoners held rights to cut turf or wood, to fish, and pasture cattle.

The 1801 Act allowed for the dividing, allotting, and enclosing of the waste grounds and commons and commonable lands within Frimley Manor in the Parish of Ash, Surrey. By 1820, the parish workhouse was established on part of the site and housed nineteen paupers. Throughout the late 1700s and early 1800s, the Frimley overseers provided for the parish poor with money, clothing, food, or fuel, funded by a Poor Rate levy on the parishioners. In some cases, if not infirm, the poor might be employed, which included physical labour like cutting turves for fuel, digging graves, or extracting stone for building. The turf was cut on the common waste much of which was unenclosed peat-moor.

At the time of enclosure, the Act stipulated that

> ....such part of the waste Lands of Frimley as in the judgement of the Commissioners was adequate to provide a reasonable supply of fuel for those inhabitants of the Hamlet who did not occupy lands or a dwelling of an annual value of more than Five Pounds.

In effect, all agricultural labourers, cottagers, and small tradesmen would qualify for fuel from the 'Firing' Allotments. Areas of land were specifically set aside for this purpose. Along with the right for those qualified under the Act to take fuel away, the Trustees were empowered to lease the whole or part of the allotments to any person they thought would be a suitable tenant. The rent was quarterly and on expiration, the tenant had to leave the land in good condition. The Trustees had spent the income on the purchase of 'fuel for firing' under the £5 qualification.

The cost of providing fuel was not too great when the bulk was wood or peat. However, as these declined and were displaced by coal, costs increased. With the situation acute by the early 1860s, there was an offer to purchase some land from the Trust and with Charity Commission permission, the sale was allowed and money invested. The Charity Commissioners also waived the interpretation of the original Act to provide fuel, so

> When not required for the purchase of fuel shall be laid out by the Trustees in the purchase of warm clothing and blankets to be distributed by the Trustees at their discretion to the deserving poor resident in the parish of Frimley.

Fuel was the priority but warm clothing and blankets might qualify. By about 1860, the poor were burning coal along with wood and peat, and more money was needed. With total income from rents and investments around £30 per year, the outlay on coal was about £1 a ton with a quarter of a ton per person per year. For clothing expenditure, a blanket was about 3s, a pair of shoes 12s, a petticoat 5s, and a pair of trousers and a man's coat £1.

By 1894, the Frimley Urban District Council was formed and after four years the Fuel Allotment Charity's Trustees passed over responsibility for the Fuel Allotments to it. By the early 1900s the Council was looking to sell the land, extinguish the rights, and purchase land for a recreation ground elsewhere. However, the Charity Commission turned down the original proposals as unacceptable, but combined with the military use of the recently enclosed common (purchased by the army), it was deemed acceptable. This allowed the maintenance of the heath and cutting of furze fuel.

Tentative approval was given in 1902 for detailed discussions with the War Office, and this agreement brought an income of around £140 per year. Along with this, the Charity Commissioners provided a 'Scheme' of guidance for the Trustees now in the Council in the administration of the Fuel Allotments Charity, with rules on how the area should be managed. In 1904, 112 people or families received the Christmas coal allocation in Camberley and Yorktown (30 tons), with 120 in Frimley, Frimley Green, and Mytchett (19 tons) – in total costing £50. By 1909, the annual income was around £200, equivalent to around £20,000 at 1990s values.

Nevertheless, there were disputes about rights to allocations, especially over 'rights' to take sand and gravel. This was hard to assess or to police effectively and caused a significant nuisance. One interesting observation was that many residents assumed a right of common even though the 'common' had long since gone and the remainder was endowed to a charitable trust. Confusion and resentment were rife particularly when the army took over. Notices of restrictions on sand and gravel getting were clearly displayed but there was nothing about rights to take firewood or to cut trees. Indeed, some people could do this by agreement with the trustees. Mr Pearce, a broom-maker, for example, could collect brushwood for his trade. Permission could be granted by the caretaker for the occasional taking of trees, with no more than two allowed each year. However, the uncontrolled and indiscriminate cutting of far more than this (up to eighty on one occasion) was causing serious damage to the site.

With the outbreak of World War One, the use of the Allotments for military training resulted in the digging of several miles of trenches and the removal of around 1,000 trees to give the impression of a French battlefield. This impact combined with that of a prisoner-of-war camp without sanitation and which generated foul smells across the area. After the war, the army compensated the Council for its losses and began re-planting. By 1939, although the price of coal had increased, the income and expenditure of the Trust remained as they were in 1909. World War Two brought further military training, more use of the Allotments, and additional damage to the now fragmented peatlands. Furthermore, with still limited income the Trust was struggling to maintain the area.

By the early 1970s, with increasing urbanisation, the Fuel Allotments were one of the few remaining open spaces in the Urban District. Alongside pressures to develop was the idea of creating a golf course over the land. However, the Charity Commissioners reminded the Council that the Allotments were for the charitable benefit of the Frimley Parish poor not for the general residents. Then, with the M3 motorway constructed close by, the population rose by 20,000 people in less than fifteen years. However, in 1973, military presence on the heath ended and the go-ahead was given to design and construct a municipally owned golf course on 210 acres of former Fuel Allotments. Objections to possible housing and loss of public open-space access meant the development proposals were shelved.

In 1979, the Council received permission to go ahead with a Local Authority-owned golf course on 120 acres with 160 acres left for free public access. There was by now a deep-seated conflict between locals who demanded the unfettered use of the area as 'common land', which it had not been since the early 1800s, and those who saw its main function as raising funds to dispense to the poor and needy of Frimley Parish. There were also those interested in the development opportunities that would inevitably arise in case planning consents were granted. In the face of these conflicts, the Council established a new charitable body to oversee the future of the Frimley Fuel Allotments. This came into being in 1983 but by 1985 it seemed the Council was unlikely to undertake the golf course proposals itself. The Trustees, therefore, decided to move independently and seek a developer interested in taking this forward. So, by 1986 there were heated public meetings to discuss the options and the future, with angry exchanges between 'dog walkers' and those interested in the charity income. From these debates, there emerged several developers wishing to pay several millions of pounds for the opportunity to develop the golf course and varying amounts of housing. Due to the delay by the Charity Commissioners of at least two years in deciding on the issues, the losses in the charitable income to be distributed to the poor of Frimley were estimated to have been around £90,000.

Eventually, after further debate and protest, permission was granted in 1988, to begin work on the new golf course. This was paid for by the developer who would also pay £100 per acre in annual rent plus 8% of the gross income from Green Fees and the Driving Range. This was with a 125-year lease with a break clause after twenty years. The result was the Pine Ridge Golf Course extending over 164 acres of former heath, leaving 98 acres to free public access. By 1994, the Frimley Fuel Allotments Charity was dispensing around £60,000 per year to the needy of Frimley Parish, and this still included the elderly needing assistance with fuel bills. This story provides a well-documented example of the military training use of a peatland site. The intensive use by the army was a key aspect in the changing landscape and the loss of the common. Ultimately though, despite the process being long and drawn out, the moors, peat bogs, and heath of this once-extensive peatland situated south-west of London has been almost totally wrecked. The detailed historical account of the fuel allotment here helps provide a remarkable insight into processes played out but hardly recorded on many other peat sites nationally and internationally.

Peatlands in the European theatre of war

Peat use was important in many countries of northwest Europe, particularly in Germany, Holland, and the other Low Countries, as well as further east into Poland. The French situation was discussed by Rotherham & McCallam (2008) and makes an interesting case study. In 1869, France exported 321,000 tonnes of peat at 10 francs 20 cents per tonne. Colomer & Lordier (1903) detail fuel usage and resources in France at the turn of the nineteenth century, but by 1914, the amount was insignificant (Berthelot, 1941). Then in the early 1940s, under the pressures of German-occupied France, there was renewed interest in the industrial exploitation of peat as fuel and for chemical products. This focus on peat-turf as fuel during times of shortage or of adversity was noted for regions such as Scotland, Canada, and the USA (Rotherham, 2005) and the situation in wartime France was a similar response to the crisis.

The inventory of French peat resources at the time suggested around 1,200,000 hectares of valley peats, marine peats, and lowland plains peats were available for exploitation. A report in *La Nature* (1941) concluded that this peat resource could help reinvigorate the French economy. However, by 1980, the annual peat production in France was 50,000 tons for fuel and 100,000 tons for horticulture; about half that of the UK.

Peatlands in warfare

As noted earlier, the main role of peat bogs and other wet landscapes in times of conflict has been in the strategy of warfare and the battlefield. However, it was especially in World War One, that bogs, marshes, fens, mud, and peat, were so obviously important. Major battlefields such as the Somme, Flanders, and others were mostly in flat or rolling countryside and dominated by bogs, rivers, marshes, and farmland. This whole landscape quickly degenerated into fields of mud as the war progressed.

Another link between war and peat has been the migration of skilled workers and the application of prisoners of war, to the reclamation of bog and fen. In the 1600s for instance, Huguenots and Walloons, escaping Catholic persecution came to England's fenlands and helped drain them. Prisoners of war from the Anglo-Dutch and Anglo-Scottish wars were used this way and many died of the wetland-borne malaria in the process. They were buried in the soft peat by the drains they dug.

The Florida Everglades – the USA's first Vietnam

In 2006, Michael Grunwald wrote a classic account of peatlands and human conflict in his book *The Swamp: The Everglades, Florida and the Politics of Paradise*. This details not only the battle with wetlands by American capitalist aspirations, values, and business but the warfare with Native American Seminoles during the early 1800s. Knowing the landscape and able to subsist deep in the swamps, the natives for many years outwitted and outfought the American military. The thankless task of eradicating the natives was described as America's 'first Vietnam War'. The removal of the Native Americans was only one part of the bigger attempt to wrest the Everglades from nature and bring them into good, capitalist production. The longer story is that the attempts largely failed and the eventual outcome was the establishment of the Everglades National Park and the celebration of the landscape and as one of America's true wilderness areas.

Further north in Virginia there are extensive peatland nature reserves including the National Wildlife Refuge site evocatively named 'The Great Dismal Swamp'. The name alone gives an insight into how European settlers viewed this peat landscape when they settled on the eastern seaboard of the USA.

The Peatlands of Holland

Peat and peatlands have been central or at least important in many human conflicts. To end this chapter, there is a further remarkable story of peat and warfare strategy. This is the recapture from the Spanish, of the massively defended city of

FIGURE 9.3 Sphagnum carpet in the English Peak District.

Breda in the Netherlands during the Eighty-Year War, and which used a Trojan Horse-type plan. The victory is still recalled by the Dutch as the '*Turfschip van Breda*'. In this case, on 25th February 1590, a turf barge with seventy soldiers hidden in the peat was left for the Spanish occupiers to take inside the fortress. The strategy was successful and just as with the 'Trojan Horse' the soldiers were able to open the city gates and the vital fortress of Breda fell to Maurice of Nassau, Prince of Orange (Rotherham, 2013a).

This short review examined case studies of the relationships between peat, peatlands, people, and conflicts or warfare throughout the centuries. The important message from this is that in these deeply eco-cultural landscapes much of what we have today has been altered, removed, or even preserved by human activities in times of stress. This is a largely unappreciated aspect of peatland conservation, but I suggest that understanding the history of how the source became what it is today is vital to any planning to repair the damage and create sustainability (Figure 9.3).

PART 4

Peatland destruction and loss

10

PEAT REMOVAL AND PEATLAND DESTRUCTION – PART 1

The lowlands

A global loss

Global peatlands have been massively reduced, in many cases so completely drained and stripped that all knowledge of their former existence has gone. This chapter sets the scene with a broad overview of worldwide peatland losses, then across North America, Europe, Asia, and in detail, in the British Isles. The destruction of tropical peatlands is also considered. Case studies are used to consider the processes and mechanisms of destruction and removal in detail along with the scale of loss. These vary over time and between uplands and lowlands, from temperate and boreal to tropical, from medieval drainage to nineteenth-century 'improvement', and twentieth-century industrial cutting.

In Europe, the process began surprisingly early with peat exploitation from before the Romans (Rotherham, 2009); though the early impacts on the overall resources were limited. However, the broad-scale removal accelerated from the medieval period onward through and into industrial times. In lowland areas of England for example, thousands of square kilometres of fen and bog were entirely destroyed between 1600 AD and 1900 AD (Rotherham, 2010, 2013b). In the uplands, vast areas of blanket mire are now drained and desiccated. Around the world, as European colonisation occurred during the 1800s and 1900s, the drainage of peatlands and the removal of peat followed (Rotherham & McCallam, 2008; Rotherham, 2014a).

Industrial air pollution adds another, often final twist of the knife as it destroys the delicate sphagnum mosses which are the architects of the peat bogs. Together with peat-cutting, and with recreational moorland 'wildfires', this demise of the bog-formers represents the latter stages of mire degradation in situations where some semblance of a peatland remains. In many cases, all evidence has long-since gone.

FIGURE 10.1 Commercial turf-cutters at Burwell in Cambridgeshire in the mid-1900s.

Recent research provides evidence that the extent of peat removal has been far greater than previously realised. This observation has major implications for understanding peat-related release of the greenhouse gas, carbon dioxide, into the atmosphere, and thus for the assessment of the role of pre-industrial carbon release in climate change. Furthermore, the dramatic scale of wetland removal (peatlands in both upland and lowland situations), has radically affected the planetary ability to moderate carbon dioxide fluxes and bring about homeostasis. On a national scale in Britain, for example, this change has affected the abilities of ecological systems to respond to climate change scenarios. Essentially, the peatland systems have been terminally compromised and the suggestion is that the scale of destruction and the implications of this are much greater than previously recognised (Figure 10.1).

A brief overview of drainage and land improvement internationally

Human impacts on the landscape through agricultural 'improvement' have occurred over the millennia; the effects on peatlands increasing over the last thousand years, and Europe a centre for developing technologies to transform peat landscapes. Understanding drivers, barriers, and actors in the social, economic, and political interactions in the countryside helps inform our vision of contemporary peatlands. Some factors forced or facilitated change, whilst others prevented it. In Europe, from around 1000 AD to the mid-1300s, the population doubled but the change was not evenly spread. Western and Central Europe, France, Germany, and England experienced population surges of around 300% which put huge pressure on natural resources (Rotherham, 2014a). At the same time the growth meant labour was available to bring about major changes.

Across Europe, internal colonisation and associated land 'clearance' or 'improvement' went hand-in-hand with population growth, with the High Middle Ages radically altering the cultural landscape. Pioneers moved into mountains to clear extensive forests, and in lowland areas marginal lands such as marshes, bogs, and fens bore the brunt of human expansion. In coastal and riverine wetlands, the changes required drainage dykes and ditches and evidence of such expansion remains in place-name suffixes; such as -bourg in Normandy, -bastide in south-west France, and -berg, -rode, and -reuth in Germany (Rotherham & McCallam, 2008). The physical layout of planned townships also reflects this human expansion. The English-'improved' landscapes in lowland areas are full of wetland and peatland prefixes and suffixes: Fenny-, Fen-, Wet-, Marsh-, -mire, -mere, -moor, etc. Some of the clearance and cultivation were done early and therefore there is little surviving evidence or awareness of the former countryside. Other areas were cleared late and retain elements of earlier landscapes. In France, for example, with the northern floodplains and the southern high plateaux, the process of internal colonisation and clearance was diverse. The Paris basin and the broad valley of the Loire, for example, were favoured zones that were colonised early. Then in the late twelfth century, there followed waves of clearance and settlement that established patterns of both rural land use and associated peasants' rights over subsequent centuries. Peat and peatlands were significant in many of these areas though today there is only limited awareness of this past.

Germanic colonisation of eastern countries was long-term and complex, and the pattern for peatlands is interesting. Settlement by skilled peasants began in earnest in the twelfth century as landowners systematically recruited colonisers who had the experience of draining and farming peatlands. This was in part in exchange for advantageous grants to the incomers of rights in the newly reclaimed lands. The same approach applied to drainage specialists imported to England from the Low Countries in the 1600s and 1700s. Similarly, Slavic princes recruited experienced peasants from long-established drained farming areas in Germany and the Netherlands. In these circumstances, skilled peasants were prepared to trek long distances from densely populated Holland, Flanders, and Westphalia to their new lands. These were mostly between the central Elbe, the Erzgebirge, and Lower Lusatia. Once they had arrived and settled, the peasants were able to apply their skills in draining peatlands and other wetlands to transform the countryside.

Nevertheless, the European population did not continue to boom, as land exhaustion, climatic deterioration, and wars took their toll. Famine increased and disease, particularly the fourteenth-century Black Death plague, drastically reduced human numbers across the continent. This collapse was followed by climatic deterioration known as the 'Little Ice Age'.

As a result, lands such as peatlands reclaimed from marginal 'wilderness' areas were now abandoned and reverted. Furthermore, coastal zones were vulnerable to sea-level rise and increasingly violent and frequent storm events. Human populations in Europe stagnated up to the fifteenth century then began a dramatic rise. So, under constant threat of food shortage and starvation there was a new phase of cultivation and land clearance. Wherever possible the transformation involved drainage and clearance of peatlands or other wetlands. However, up to this time such changes were limited by the available technology. Deep bogs and fens were just too difficult to tackle; besides, many local communities depended on these lands for their subsistence. This meant there was significant resistance to change.

There was some pressure on the lands previously cultivated but subsequently abandoned because of Black Death. However, population growth meant increased exploitation of previously uncultivated lands in mountains, woods, marshes, fens, bogs, and along rivers and coasts. Wherever possible, moors and ancient barren and unproductive grounds were cultivated, meadows turned to arable land, and forests destroyed. One compounding factor was that landowners tried to limit forest destruction (to safeguard vital resources) and this meant peasants exploited other areas such as peatlands.

Continuing drainage and exploitation of bogs and marshes occurred along the North Sea coastline, with extensive drains and dikes. In the Netherlands, around 44,000 hectares of marsh and fen were reclaimed from 1565 to 1615. The German coast had 40,000 hectares drained and protected during the late 1500s and early 1600s, and peatlands in Eastern Europe were similar. Lithuanian peasants migrated to the north of East Prussia, and Dutch Mennonites abandoned the Netherlands for religious reasons and colonised the Vistula-Nogat Delta from about 1550. Here they exerted their skills to drain the bogs and fens, and Polish manorial lords were so impressed by the resulting increases in land values, that they sought Dutch drainers whenever they could (Rotherham & McCallam, 2008).

Tensions often grew between old, established rural communities and the incomers over colonisation and improvement of peatlands. This was especially so in the English Fens when European settlers arrived to drain their wetlands (Rotherham, 2013b). Problems arose over issues such as common rights which in marginal landscapes were hugely important. Competition over rights and resentment of the drainage schemes soon led to violence. With population increase and controls over common usage, came issues of 'ownership' and protection for commoners who were already in place. This pressure of human need, unrest, and sheer numbers of people pushed nature and the environment to an irreversible tipping point. In relationships between agriculture and natural resources the development of industry is critical. Industry and urbanisation need increased agricultural production to meet the rising demand for food. Agriculture also has to 'release' resources (labour, capital, land) which transfer to industry and services. Agricultural revolution goes hand-in-hand with industrial evolution and frequently with political revolution too.

Changes that brought about agricultural revolution included new techniques and equipment to cultivate soil, new ways to drain and 'improve' the land, advanced crop rotations, heavier stocking of animals, and the intensification of mixed husbandry. The processes included increased productivity on land already farmed and the extension of agriculture into new or long-since abandoned areas. After around 1840, there was also the option of increased artificial chemical fertilisers and rapidly evolving farm machinery as science improved technology and the knowledge of chemistry too. Ways in which land was held, owned, or managed, changed with parliamentary enclosures and the associated consolidation of farm holdings.

Peatland destruction in the British Isles

I am not going to attempt a definitive account of draining since there is enough on that to fill many volumes. Each fen, each marsh, and every common has its own unique story. Instead, this is a broad overview of what was in many ways a simple and straightforward sequence of events but which in practice was also horribly complicated. The real process amounts to a hugely complex mess of politics, history, environmental calamity, and local twists and turns. There are overall general 'truths' about the process of land improvement and drainage which occur repeatedly throughout history. The gradual incursions of people into the great wetlands and peatlands occurred over many centuries but when it comes, the end is speedy. Nevertheless, the landscape often fluxes as nature often recovers the land to drive human colonisers back. Eventually, though, the peatlands succumb to human attrition.

The final act of drainage is often not the end of the process but one major step along the way. In both the Northern and Southern Fens of England, following drainage, there were intensification and desiccation of the wider countryside of drained fens and bogs. In England's fens the main period of drainage was from the mid-1600s to the mid-1800s, beginning perhaps with the Humberhead Levels and the Great Level and ending with the obliteration of the once-great Whittlesey Mere of Cambridgeshire. Eventually, the final remnants of the fens were mopped up in the 1940s war efforts to feed the nation. In the wider context though, the process of removal began well before the 1600s and continued after the 1940s.

The example of the English Fens shares common history, exchanged cultures, communities, ideas, and technologies with European peatlands and global wetlands generally. 'Improvers' travelled the world in search of peatlands and other wetlands to drain. These processes of peatland drainage and land improvement do not stand alone but are influenced by politics, religion, and science or technology. Furthermore, the changing peatlands represent just one part of a broader landscape change throughout the 1700s and 1800s movement for scientific improvement.

European agriculture and land use were influenced by the writings of natural philosophers of the enlightenment. Agricultural improvers like the Englishman, Arthur Young (e.g. Young, 1804 for example), helped revolutionise attitudes to farming and land with the emergence of enlightenment and rational thought.

Old and traditional values were pushed aside as a new zeal for 'land improvement' developed. Some of the resulting impacts became cataclysmic such as in England where civil war followed the 1600s attempts at peatland drainage. This countryside also fostered both Methodism and Quakerism which emerged as radical and threatening ideas during times of philosophical conformism.

Exchange of ideas and methods of land improvement were triggered by the changes. So, following exile after the Civil War, Royalists coming home to English estates brought new ideas learned in the European Low Countries. Armed with ideas of land improvement from Holland, Flanders, and France these landowners also needed to repair their financial situations and saw peatland reclamation as one way forwards. The Third Earl of Winchilsea, Heneage Finch, had left his Yorkshire estate in the East Riding after the Royalist 1655 uprising and returned to debts from his role in this. He arrived home from Europe with ideas for changes to improve farming and increase rents and these focused on his 'low lands', that is, peat fen and carr. The intention was to deepen the beck with ditching and draining of the extensive wetlands. A portable horse-mill was to pump out the water and then large areas were to be planted with alders and willows. This reclaimed landscape would be stocked with cattle and horses (Rotherham, 2010).

In the process of the exchanges in skills, knowledge, and especially of finance to bring about change, the Dutchman, Cornelius Vermuyden, was to prove a pivotal figure. He moved to England in the early 1600s and became the first notable 'drainage engineer' to attempt to drain extensive peatlands. Vermuyden's first large-scale commission was the draining of Yorkshire's Hatfield Chase from 1626 to 1629. He went on to put forward similar projects to drain the Somerset Levels for example, but these never came about. The engineering operations in the English Fens and for example, the French coastal marshes were proto-capitalist exercises "directed by an association of technical experts and business men [...] financed by a few large business-houses, mostly Dutch". The aspirations combined public land reclamation and private wealth generation. As such these initiatives became the economic models for later eighteenth-century reclamation enterprises across Britain and Europe. For instance, in Brittany and Guienne in France, "companies were founded for the express purpose of financing – or indeed speculating in – land reclamation, which now also received government patronage" (Rotherham & McCallam, 2008).

However, the drive to remove peatlands from the landscape was not simply drainage and improvement for financial gain. Peatlands like marshes, fens, and bogs were known as places of evil, infamy, and as such these challenged religious and public morality. They needed to be 'reclaimed' and the idea of 'reclamation', and later of 'improvement', resonated with strong moral ideals. The first engineers and drainage experts in England were frequently Protestants and Huguenots fleeing continental religious persecution and civil strife. With powerful religious undertones of the threat of 'biblical' flooding in bogs, marshes, and fens, this drainage was addressed with religious fervour.

11

RECLAIMING THE PEATLANDS

Draining the land

In Britain and then in continental Europe, by the late 1700s, there were developments in land-drainage techniques which improved attempts at peatland reclamation. Around 1764, the improver and farmer Joseph Elkington solved some vital problems concerning under-drainage. This meant surface waters from a wetland could be removed at the same time as siphoning off low-level underground water tables by tapping and diverting springs. To give some acknowledgement of this achievement, at the time, the British Parliament in 1795, awarded Elkington £1,000 in recognition of his innovation and its contribution to land improvement (Elkington, 1797) (Figure 11.1).

A major incentive for eighteenth-century English peatland drainage was the development of enhanced food production to supply growing urban industrial populations. Arthur Young, the agricultural 'improver' and writer, in 1772, stated that, "breaking up uncultivated lands" and "draining fens" were the main means for enhancing productivity and raising the incomes of those farming them. This logic was simple since reclaimed peatlands might be enclosed and incorporated into new crop rotation cycles without fallow. The improved peatland could grow fodder crops for livestock and these animals in turn generated manure to fertilize the reclaimed land.

Draining the English fenlands

Peatland drainage in the lowlands transformed landscapes which were swept away by a zeal for improvement. "Through their efforts in wresting the land from the swamps the refugees bequeathed to our generation rich alluvial soil by which the Fens earned an enviable title: The land of Goshen [i.e. The Land of Plenty]" Bevis (1983).

Over this period ideas and knowledge about the drainage and subsequent of peatlands were exchanged across the wider landscape of Europe. Dutch drainage

FIGURE 11.1 Bog-moss was once widespread across lowland bogs; this shows a sphagnum carpet at Thorne Moors, England.

engineers came to England to transform the low-lying peatlands of fens, marshes, and bogs. 'Improvement' throughout the 1700s and 1800s meant drainage and destruction of peatlands on a grand scale. In an historic context the transformations were swift. The English Great Level of the Fens in Cambridgeshire remained relatively intact, a huge, wet peatland until the late 1500s but was gone by late 1800s. By the late medieval period with climatic deterioration there were constant threats of inundations and damage to property, settlements, and areas that had earlier been reclaimed from the peatlands. The drainage engineers, Smiles, observed in 1904, that

> ....it would be difficult to imagine anything more dismal than the aspect which the Great level then presented. In winter, a sea without waves; in summer, a dreary mud-swamp. The atmosphere was heavy with pestilential vapours, and swarmed with insects. The meres and pools were, however, rich in fish and wildfowl. The Welland was noted for sticklebacks, a little fish about two inches long, which appeared in dense shoals near Spalding every seventh or eighth year, and used to be sold during the season at a halfpenny a bushel, for field manure.

The great drainage begins

By early 1600s, soon after James I assumed the throne of England, there was a series of very damaging floods that burst the defensive flood embankments along the county's eastern coastline. This disaster was associated it seemed with long-term neglect of flood protection and, therefore, the King was under pressure to take positive action. It was stated that,

> For the honour of his kingdom, he would not any longer suffer these countries to be abandoned to the will of the waters, nor to let them lie waste and unprofitable; and that if no one else would undertake their drainage, he himself would become their undertaker.
>
> *(In Wentworth-Day, 1954)*

The result was that a Commission was established to report on the problems and they found that at the time, more than 317,242 acres of land requiring draining and protection was outside defensive dykes and drains. However, in the absence of specialist engineering skills, the early attempts at control were modest and often unsuccessful. Despite the limited early success, this was a tipping point and signalled the beginning of the end for many lowland peatlands across England. On the back of the initial attempts and their limited success, the Dutch 'engineer', Cornelius Vermuyden, was invited to propose schemes in England. He was known to have had success in ditching, draining, and embanking areas in the Low Countries. The King, anxious to drain and reclaim Crown estates at Hatfield Chase in Yorkshire, commissioned Vermuyden to oversee the necessary work. Shortly before this, when Henry VIII's antiquary John Leland, visited Hatfield in the mid-1500s, probably around 1534, he had travelled through the area by boat. Yet within a few years this peatland was dry, fertile farmland (Rotherham, 2010, 2013b).

In 1600 AD, the English Parliament passed "An act for the recovery and inning of drowned and surrounded grounds and the draining dry of watery marshes, fens, bogs, moors and other grounds of like nature". Whilst this was primarily directed at the East Anglian Fenlands, it was first used for the Yorkshire Fens around Hatfield Chase. After a visit by a royal hunting party, which included Vermuyden, in 1607, the 1608 survey of Royal Forests and Chases described Hatfield Chase as 'utterly wasted'. It was a rich hunting ground but a poor cropland as the clock was now ticking towards its fate. Indeed, the act of parliament spelt the end for many lowland peat sites. The great drainage began with the peatlands in Yorkshire and North Lincolnshire and then swept through those of East Anglia. The once-massive mire system situated between Norfolk and Suffolk had already succumbed to extensive medieval peat-cutting. The resulting former peat-working then became flooded ultimately becoming the Norfolk Broads.

As the drainage progressed, though still needing long-term cultivation, the now dry reclaimed land became available for agriculture. Impressed by his work, in 1629, King Charles I presented Vermuyden with a knighthood. This award was in "recognition of the skill and energy that he had displayed in adding so large a tract to the cultivable lands of England". However, the early improvements were not as impressive as they initially appeared and the new drainage systems triggered major problems in the previously farmed countryside. Furthermore, there were tensions and violence breaking out between locals and Dutch incomers, and Vermuyden and his workforce abandoned South Yorkshire for the East Anglian southern fens. Damage to the 'old lands' caused by Vermuyden's project was eventually put right, but this was at great cost. It was necessary to cut a major new drain, the 'Dutch River' which took the River Don north from its confluence with the River Trent in the south and to the River Ouse in the north. There were difficulties in raising the necessary finance for the additional works since the earlier problems had lowered the value of shares in the 'adventure'. This was also a time of major social and political unrest and, triggered in part by plans to drain the southern Cambridgeshire peatlands, the English Civil War erupted across the nation.

Whilst some of these early attempts at mass drainage were undoubtedly successful, there were some basic and intractable problems. In draining the peatlands there was the issue that once dry, they shrank and the ground surface dropped. Dry peat dissipated by wind-blow and erosion, broke down and released carbon dioxide into the atmosphere. This soon meant that drained peatlands became lower that the drains and the flow of water could reverse and back up. Furthermore, in the absence of the necessary technology to pump the water out against gravity the drained peatlands became wet again. Over the following two centuries there were piecemeal schemes to drain and dry the countryside turning extensive peatlands into intensive farmland.

Removing water from the land was a big problem and windmill pumps were employed to keep the lower-lying peatlands dry, which worked as long as the wind blew. However, because the wetlands were in coastal zones, there were the dual problems of keeping seawater out whilst at the same time discharging freshwater into the sea. This situation presented a potential contradiction with sluggish rivers and little 'fall' across the landscape. The rivers which drained the fenlands had a tendency to deposit sediment-rich loads at the river-mouth and the estuaries then blocked. The result was that the freshwaters backed up and caused inland flooding.

However, the drainage was not yet complete and what followed over many decades were 'improvements' on individual fens and other peatlands across Britain and Europe. Then, within the drained landscapes now turned into enclosed farming fields, further ditches and under-field drains removed water on a micro-scale. Rivers now flowed through artificial banks into new channels with new dykes, drains, cuts, channels, new embankments, and sluices. Channels would be dredged and deepened, and pumps took water from wet areas into ditches and drains, and then from side-drains into main-drains, and away. The early pumps were windmills but these were followed by steam pumps, oil pumps, and finally electric pumps as technology and engineering gave humanity the upper hand over raw nature. Sluices kept the freshwater from backing up and prevented seawater from breaching into low-lying peatlands. Great embankments were constructed to line the 'natural' rivers and the constructed channels, and they also protected coastal zones from high tides and storm events. In Britain just in the Southern Fens alone, it was calculated that around 680,000 acres of peatland were converted from 'dreary waste' into the most fertile and productive farmland (Smiles, 1904).

> The soil of England is sufficiently fertile; but it has been made so principally by the industry of its people. A large portion of it, and that the most fertile, has been reclaimed and embanked from the sea. The two thousand square miles of land in Lincolnshire and Cambridgeshire, forming the Great Level of the Fens, would, but for human skill and industry, be lying under water, the haunt of wild birds.

Smiles was enthusiastic about the results,

> The prophecies of the decay that would fall upon the country, if 'the valuable race of Fenmen' were deprived of their pools for pike, and fish, and wild-fowl, have long since been exploded. The population has grown in numbers, in health, and in comfort, with the progress of drainage and reclamation. The fens are no longer the lurking places of disease, but as salubrious as any other parts of England. Dreary swamps are supplanted by pleasant pastures, and the haunts of pike and wild-fowl have become the habitations of industrious farmers and husbandmen. Even Whittlesea Mere and Ramsey Mere, – the only two lakes, as we were told in the geography books of our younger days, to be found in the south of England, – have been blotted out of the map, for they have been drained by the engineer, and are now covered with smiling farms and pleasant homesteads.

This was a eulogy on the demise of the region's peatlands and provides just one example of a process which inexorably rolled out across Britain and Europe.

The landscape-scale early projects were followed by 'mopping-up' operations focused on enclosure and drainage awards for individual commons and implemented by particular drainage engineers. Unlike the early self-taught 'amateur' drainers, these were professionally trained engineers. A notable individual was John Smeaton from Leeds who in the 1770s worked up a scheme to drain the largest of the South Yorkshire carrs, Potteric Carr. Smeaton drained Holderness in 1763, working with engineer John Grundy to improve drainage for around 11,000 acres of low ground, that is, peatland in the Hull Valley. He worked on Hatfield Chase in 1764 and on Adlingfleet Level near the River Trent in 1772. The latter scheme involved comprehensive drainage taking floodwater off about 5,000 acres of low ground for a local landowner. In England these works might be undertaken piecemeal fashion but often required collaboration by major landowners and parliamentary acts to provide consent for larger-scale operations (Rotherham, 2010, 2013b).

The task of draining and improving the great peatlands was a major undertaking with formidable challenges. One of the most difficult of these was the draining of vast flood lands and raised bogs which formed the core of the great Yorkshire Fen. This task ranks as one of the major engineering feats of the time and involved activities spread out over the wider fens and peatlands and along each arterial river valley. By around 1637, the Ancholme Valley to the south-east of the fen had a drain cut to the Humber estuary and a tidal sluice constructed to control the seawater. Nearby areas of carr peatlands were successfully drained though the land further north still flooded.

Extensive Holderness peatlands of fen and carr were under threat and by the nineteenth century, the Beverly-Barmston Drain and Holderness Drain were dug and the River Hull was dredged. Although the area was now drying out with piecemeal drainage since medieval times, there were still post-glacial meres that were valued for fowling and fishing. Many of these smaller peat sites had been drained by monastic improvements in the 1200s and 1300s, though most were lost in the 1500s and 1600s. By the late nineteenth century, remaining site of significance was Hornsea Mere and entire peatlands had been simply erased from maps and even from memory.

The final stages of removal were the 'mopping-up' operations of remaining wetlands which continued well into the 1970s. Alongside drainage and protection from flooding of historic peatlands were the increased enclosure and 'improvement' in the wider countryside. In Britain this was undertaken through individual acts of parliament to allow owners to enclose, improve, and parcel up the countryside into modern farm holdings. Lands owned by a few local people but with common rights attached for the wider community had the rights largely extinguished and the land was subsequently drained, improved, and redistributed amongst the select few. Drainage of wetlands and under-draining of relatively drier lands were at the heart of these operations. The process meant that over a period of maybe two centuries farmland was taken from fen, heath, bog, marsh, or 'waste', and there was a progressive, insidious intensification of farming operations to create the 'bread-basket' landscapes of the twenty-first century. In the fenlands of east England, almost all the once-extensive peatlands were drained and improved, and the drier grounds around were also enclosed and farming intensified. Following the local landscape histories though the centuries we see how the unimproved landscape with its vast peatlands was converted and transformed into the modern farming countryside of today. The peat sites which remain are generally much smaller, drier, and fragmented than the huge wetlands of the medieval period. Furthermore, as sites desiccated the peats shrank, and also, many were cut for fuel. This example from eastern England demonstrates the process and the timeline, and this was acted out on peatlands across the world over the same period.

Impacts of draining

Soon after the earliest attempts at major drainage the problems of peat shrinkage were showing, even though the actual causes were not understood. The most well-known example today is that in Holme Fen in Cambridgeshire. The local squire, William Wells, had been involved in the drainage of the great Whittlesey Mere in the mid-1800s. From his observations elsewhere in the Fens, he quickly realised the likely scale of the resulting ground shrinkage. With this in mind he set up the now-famous Holme Fen Post to indicate the scale of the impact. Originally the post was buried with its top level with the ground surface and by the late twentieth century it was four metres proud of the now shrunken peat. This loss provides a starting point to calculate the carbon loss to the atmosphere from the wider fen landscape and contributing to processes such as climate change. Furthermore, the countryside of the former Whittlesey Mere is now the lowest-lying land in Cambridgeshire and even perhaps in England. In these shrunken peatscapes and similar areas elsewhere there is today productive, intensive agriculture but this is below sea level and requires active pumping of water to maintain it. Furthermore, as the water tables fall and the lands surface shrinks, coastal zones may become vulnerable to saline creep and hence to seawater contamination.

In England, 'The Great Draining' ended before the twentieth century but this was not the end of the process. Intractable sites clung tenaciously to life until the 1940s but then succumbed to American technology of tracked vehicles and tractors. Then there followed the excesses of government-funded, intensive agriculture of the post-World War Two years. The lid on the coffin of the old fenland countryside of bogs, mires, fens, heaths, and marshes was well and truly nailed down. This pattern was repeated across wide landscapes of Great Britain and Europe, and the ideas and processes were exported around the world too. This was not only in the low-lying fenlands but the uplands too. Moors and bogs were progressively gripped, ploughed, limed, fertilized, and converted. Some peatlands were enclosed to pasture and others were burnt and systematically overgrazed (Figure 11.2).

FIGURE 11.2 Invasive rhododendron, sometimes a major problem on desiccated peat bogs, and here shown in the English Peak District.

12

PEAT REMOVAL AND DESTRUCTION – PART 2

The British uplands

Intensive land management, extensive forestry planting, and particularly associated drainage, grew throughout the twentieth century. These were compounded by intensive grazing by livestock, and industrial-scale grouse shooting (Figure 12.1). The impact of upland the fuel economy, with widespread cutting of peat and turf, and again, associated drainage of hillside, hilltop, and valley-bottom mires, shaped much of the present landscape (Rotherham, 1999a, b; Rotherham et al., 2004). These impacts were exacerbated, by widespread cutting of ling and gorse, and gathering kindling and other materials for fuel, bedding, and fodder. Birch and bracken were harvested for a diversity of uses, including fuel and building materials. Both turf and ling were widely used as cheap materials for the construction of buildings for people and stock. These and other uses occurred over many centuries, largely on a localised, domestic scale, but sometimes also on an industrial one (Ardron et al., 1996, 1997, and 1998); the effects on vegetation, soil, and upland hydrology were massive, yet often overlooked by researchers and managers.

This oversight is for two reasons:

1. Progressive ending of many of these land-use practices, during the twentieth century, so that many researchers are simply unaware that they may have occurred.
2. The wholesale nature of some of the exploitation, to the extent that almost all direct evidence of use has been removed. For most lowland areas, intensive land use has obliterated almost all the evidence from the landscape.

Peatland extent and the example of the English Peak District

The boundary of the high Peak District heather moors has long been known to have fluctuated dramatically over the centuries. In the White Peak, almost the

FIGURE 12.1 Local domestic peat-cutting removed vast quantities of peat for fuel. Here we see a Stornoway croft and its peat-stack around 1919.

entire cover of limestone heath was lost in the 1700s and 1800s. Furthermore, there has been little attention to the moors and peatlands around the periphery of the uplands, and the extent of the former outliers to both east and west. The massive contraction away from the lowlands and into the uplands in recent centuries has been ignored. Periods of agricultural intensification were followed by abandonment and recolonisation of moorland (Parry, 1977).

Evidence of major peat bogs, moors, and heaths extends many kilometres down towards the lowlands. In the east, for example, bogs and peat utilisation are recorded from the western suburbs of Sheffield and Holmesfield and then down to valley mires and marshes of the Don and Rother Valleys. Maps produced from detailed archival research by Gordon Scurfield (pers. comm. and in Rotherham, 2011a, b) over a thirty-year period, present a unique insight into the former extent and connectivity of unimproved lands. Impacts of the changes include catastrophic losses of biodiversity and both fragmentation and desiccation of the remaining habitats (Rotherham, 2011a, b).

Upland peatlands and their cultural use

'Peatlands' are fragile environments ranging from relatively productive, lowland fens to the low-nutrient, unproductive peat bogs, moorlands, and heaths. Not all are true 'peatlands', but for this discussion they have substantial elements in common. As noted they include lowland fens and mires, upland blanket bogs, upland and lowland raised bogs, other bogs, upland wet heather moorland, upland dry heather moorland, lowland heath, grass moor, and grass and heath commons. Often described as 'wastes' and 'commons' by topographers and landowners, and 'relatively' unproductive, across much of the countryside, they provided fuel, food, and building materials for local people and sustainable grazing for domestic stock over many centuries. In upland landscapes, especially for subsistence communities, the peatlands were vital parts of the rural economy. Before enclosure and 'improvement', or abandonment, caused widespread and almost total destruction of commons, heaths, bogs, and peripheral moorlands, most people relied on them for a diversity of products (Table 12.1). Furthermore, the upland peat landscapes merge into northern and western countryside areas in particular, and much of the human cultural impacts and usage would have applied to unenclosed areas in many lowland areas too. This changed dramatically with the imposition of parliamentary enclosures in particular.

The evidence of rights to cut peat-turf for fuel is far more widespread than generally recognised. Many urban areas that now seem unlikely candidates for peat usage held traditional rights and these were used. Settlements like Sheffield,

TABLE 12.1 Traditional products from upland and other unenclosed peat landscapes

Fuel:	• Peat • Turf • Ling • Sedge • Bracken • Gorse/furze • Kindling • Birch coppice//brushwood • Rushes for lights
Grazing:	• Sheep • Cattle • Ponies/horses • Deer • Rabbits • Moss hay
Building materials:	• Peat • Turf • Ling • Stone • Sphagnum moss • Bracken or fern • Rush • Clay • Birch poles • Other wood

Other foods, medicine/herbs/dyes gathered etc.

York, Leeds, Liverpool, Southampton, Carlisle, Kendal, and many others relied on peat fuel in some cases from Roman times and up to and through the medieval period. According to the Ecclesall Enclosure Act (1779) for the land south-west of Sheffield, "…….there are within the Manor and Township of Ecclesall in the Parish of Sheffield several open commons, moors and waste grounds containing in the whole, 1000 acres or thereabouts". In this context, Carolus Paulus (1927) noted that in the books of the Vicar of Sheffield, it was stated that in 1775, Ecclesall had 1,128 families mostly described as cottagers, labourers, farm servants, and squatters. The cottagers owned or occupied cottages with attached ancient customs for the rights of commonage on the wastes. These included the right to pasture one or more animals on the common, and significantly to cut turf and extract fuel. Through research into historic rights we can piece together an image of the countryside with much more extensive peatland than exists today.

Peat removal – the scale of cultural utilisation

The scale and extent of this landscape utilisation is remarkable and includes peat-cutting, 'turf-getting', 'moss-gathering', bracken and rush harvesting, and many other uses. For upland landscapes especially, this was far greater than previously recognised. Obviously, in many lowland areas the evidence was pretty much swept away by drainage, conversion, and intensive agriculture. However, in many upland and western areas the evidence remains more intact. The extent of well-defined peat-cuttings around areas of Peak District blanket mire, in the English Lakes (e.g. Marshall & Davies-Shiel, 1971; Denyer, 1991; Rollinson, 1996; Gambles, 1997), in Cornwall (e.g. Hamilton-Jenkins, 1932), or Devon, or around the North Yorkshire Moors, for example, is considerable (Hartley & Ingilby, 1968, 1972 for example). Along with this, thin peats and turf, on low-level moors were stripped away. In many areas, there has been extensive 'turf-cutting' and 'paring and burning' for agricultural improvement. In some cases, this was associated with the creation of in-bye land.

The relative amounts of in-bye farmland resulting from the removal of turf, thin peat, deep blanket mire, or 'paring and burning' are unknown, but the collective extent is considerable. This impact extended downhill into the extensive lowlands, but in most areas the evidence was removed by enclosures and subsequent agricultural intensification. In intact areas of unenclosed heath, royal forest, chase, and the like, much evidence remains and this is supported by archival information and even by oral histories.

Shallow peat-cutting, 'turf-cutting', and 'paring and burning' took place extensively over the upland landscape, including valley-sides, and down to areas of cutting and utilisation in the lowlands. This usage affected valley mires, raised bogs, fens, and importantly, lowland heath and common. Where suitable landscape remains intact, upland or lowland, the widespread occurrence of drainage features and other cutting infrastructure, indicate that few 'peatlands' (if any) were unaffected. Upland areas such as the English Pennines (particularly the Peak District), the Yorkshire Dales, the North Yorkshire Moors, the Lake District, Cornwall, Devon, much of Wales, and large areas of Scotland experienced dramatic landscape change through fuel exploitation for peat and turf (Rotherham, 2008b, 2009).

For the Peak District and South Pennines, as demonstrated by Ardron et al. (1998), there was more medieval peat-cut (*c.*34 million cubic metres), than from the Norfolk Broads at the same time (Ardron, 1999). Mostly this was in the early medieval period when lower-lying sites were exploited and then progressively destroyed during the enclosures of the sixteenth, seventeenth, and eighteenth centuries.

Past trends and their impacts and implications

Over many centuries, exploitation of landscape fuel resources caused major changes; upland across Britain was shaped and manipulated by fuel-associated management; drainage and removal of deep organic matter changed its water-holding capacity and behaviour, with huge impacts in upland areas and lower-level catchments downstream. As noted, low-lying areas in the Humberhead Levels, the Cambridgeshire Fens, and the Somerset Levels, for example, had vast quantities of peat and turf removed and changed their ecology and landscape. Nevertheless, as long as fuel-demand and other subsistence countryside uses persisted, large areas were maintained as peat moor and fen.

Around the Peak District and the Pennines, there was massive loss of heathland and low-lying moor, and major incursions into upland zones too. This was particularly from around 1600 AD up to the present day, and the remaining unenclosed high moorland landscape has been subjected to major changes too. Most of the peat was removed in early medieval times but some exploitation continued well into the twentieth century. Today the bulk of the remaining moor and bog is in the uplands and the west of the country but this was not so historically. Any remaining lowland or southern sites are generally small and highly fragmentary with exceptions being the New Forest in Hampshire and the Dorset heaths, for example.

Peatland dynamics

Across Britain for instance, peatland landscapes continue to change but are no longer dominated by active subsistence or commercial utilisation peat-cutting. Contemporary moorland, bog, and heathland uses like sheep-grazing, grouse-shooting, and recreation, dominate many upland zones. The vegetation on the remaining blanket mires and wet moors still changes because of drains and cuttings affecting hydrology and atmospheric fallout which influences soil conditions. Historic peat pits, linear cuttings, and over-cut areas of peatland may now be distinctive and sometimes ecologically important holding uncommon plants and uncommon insects such as Odonata.

Conservation management issues in peatland environments

Extensive upland peat-cutting has been recognised relatively recently. Additionally, the former extent of peatlands in the UK has been dramatically under-appreciated. There is a further dilemma for conservation in terms of the heritage and archaeological significance of some of the ancient hand-cut peatlands. Whilst there has been some recognition of the importance of this (in the lowlands in the Somerset Levels and at Fenn's and Whixal Mosses in Shropshire, and in the uplands on Exmoor, for example), such awareness is very much in its infancy. The late Oliver Rackham (pers. comm.) described the discovery of the rich archaeological heritage of hand-cut peatlands in upland England as being on a similar scale of conservation significance to that of ancient coppice woods (Rackham, 1986).

The peat resource and the peat-cut landscape

The exploitation of peat has been widespread, involving shallow as well as deep deposits, and the peat cover has often been more-or-less completely removed (Rotherham, 2009). Because of this, the remaining peat resource is reduced considerably more than was formerly recognised. Furthermore, historic and ecological records preserved in remaining peat deposits, even if degraded, are of great scientific interest and value. In this context, shallower blanket peat and topogenous deposits on middle- and lower-level moorlands are now very scarce resources and there is an urgent need to protect as many peatlands as possible and to actively conserve to prevent further deterioration.

On upland moors, further deterioration of remaining mires can and is being checked by blocking still-functioning drains. Along with blocking and breaking of existing drains and grips, no new drains should be cut and existing drains should not be cleared. Since medieval and other peat-cutting upland landscapes have remained largely unrecognised, they have no specific protection. Up to now, any protection has been largely incidental, and because land improvement in these marginal areas has been relatively ineffective. Land uses like sheep-grazing and grouse-shooting have been favoured by owners, and serve to protect this heritage at least to an extent.

Environmental consequences of change – eutrophication, succession, and species loss

Peatlands on upland and other marginal areas can be affected by cultural severance from long-term subsistence exploitation, that is, the displacement of traditional community use by, for example, grouse-shooting and intensive sheep-grazing. Whilst some changes can be addressed through actions like the reinstatement of traditional grazing with cattle, others remain intractable but may provide opportunities for rewilding initiatives. Long-term exploitation for

FIGURE 13.1 North American peat-moss harvest stored in sacks in the 1960s.

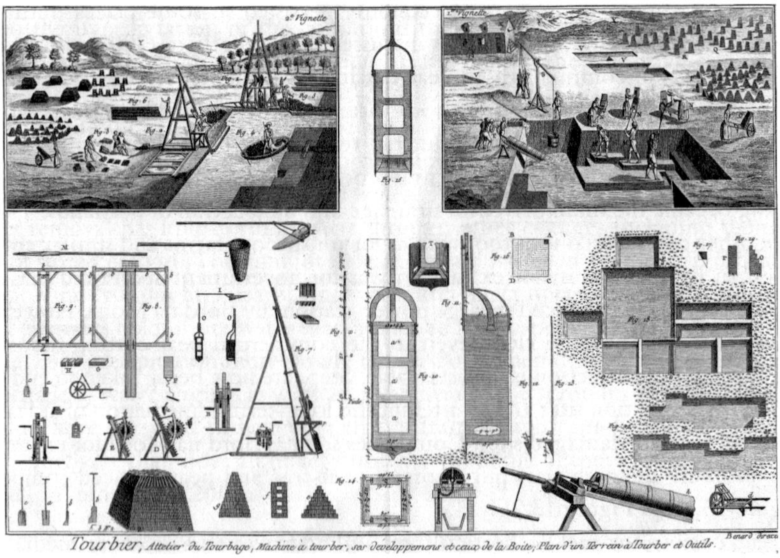

FIGURE 13.2 An illustration to show the technology and methods of French peat-cutting around 1769.

chemical works to extract and distil organic chemicals, to horse-litter, to paper, and even thermal underwear. This chapter closes with a look at some of these forgotten uses.

The uses of peat and peatlands

The potential to exploit bogs and fens to use peat taxed the minds of scientists, industrialists, and farmers for generations. This appeared to be a widespread and extensive resource that was under-utilised. Some were enthusiastic and others less so. According to Robert Alloway in 1869,

> few are now disposed to believe in the practicality of the profitable use of peat but on the contrary are more inclined to consider with the London Times that the one huge succession of bogs which fill up a large portion of the central parts of Ireland, are useless and unsightly excrescences, growing over and covering up three million acres of good land.

Yet for many ordinary people these were productive landscapes, and before their almost total destruction through enclosure, 'improvement', or abandonment, most people especially the rural poor relied on them for a diversity of products (e.g. Rotherham, 2005):

a Fuel: Peat, turf, ling, gorse/furze, kindling, birch coppice/brushwood,
b Building Materials: Peat, turf, ling, stone, gravel and sand, sphagnum moss, bracken or fern, Rush, Clay, Birch poles, other wood
c Grazing: Sheep, cattle, ponies/horses, deer, rabbits
d Other foods and herbs gathered

The majority of people in medieval and early industrial times were labourers, farm servants, squatters, and cottagers; the latter owning or occupying cottages, with ancient customary rights of commonage on the wastes. These rights were of various kinds and included the right to pasture one or more animals on the common, to cut turf, and to extract fuel. Massive conversion of heath, moor, waste, and marsh, to arable and enclosed pasture, dramatically ended this utilisation of the landscape by most of the rural poor and poorer commoners. Along with wood mostly from 'woodland' and wooded common, peat, turf, gorse (or furze), broom, bracken (fern or brake), flag (peat turf), reed, sedge, and ling were harvested from heath, common, bog, and fen. However, as stated by Pinchbeck (1930), in England, ".... turf and furze were the common fuel used by a great proportion of the population".

If it occurred at or close to the surface then mineral coal might also be taken but this was not generally favoured because most coals generated noxious fumes when burnt in a primitive way. Fuels might be combined as need and availability dictated. Gorse or furze was often an alternative to wood or peat if they were in short supply, or else as kindling or high-temperature baking.

The former extent, cultural history, and deep-rooted significance of the peat-cutting landscapes are now mostly forgotten. Much of this countryside was used for peat-cutting, 'turf-getting', 'moss-gathering', bracken- and rush-harvesting, and other applications. For upland peat landscapes especially, and for lowland heaths, bogs, and fens when they survive, this was far greater than previously realised. Around the remaining areas of blanket mire on Dartmoor, Exmoor, Bodmin Moor, the Peak District, the North York Moors, and similar areas are extensive, well-defined, peat-cuttings. Thin peats and turf on low-level moors were stripped away by extensive 'turf-cutting' and also 'paring and burning'.

For many local communities, when conversion to arable and enclosed pasture of heath, moor, bog, fen, waste, and marsh, dramatically ended this usage by the rural poor, the loss of these resources was a devastating blow. Small-scale exploitation lingered on in remote areas, particularly the uplands and the west of Britain. But by the late 1900s, almost all subsistence exploitation in England and Wales had died out and traditional use in Scotland and Ireland was rapidly declining. Commercial cutting has largely ended in England but has accelerated in Scotland and Ireland. Even when landscapes have changed and use stopped there is still a legacy of place-names and surnames:

a **Place-name evidence:** Peat Pits, Peat Lane, Peat Road, Turf Field, Moss Crop, Moss Side, Turvin Moor, Turf Moor, Turf Hill, Turbary Lane, Peatfield
b **Surnames:** Peatman, Peat, Moss, Mossman, Peatfield, Peate

Peat as fuel for the common people

Across much of the British Isles and commonly in parts of Europe, cutting for fuel was the most widespread use of peat and occurred almost everywhere that held suitable deposits. Peat and turf were not the only fuels from heath and common and use of materials varied from site to site with ownership and rights, and particular needs. The diarist John Evelyn (1729), writing about 'fuel' noted:

> … But besides the Dung of Beasts, and the Peat and Turf which we may find in our ouzy Lands and heathy Commons) for their Chimneys, Cowsheds, etc. they make use of Stoves, both portable and standing …. In many Places (where Fuel is Scarce) poor people spread Fern and Straw inn the Ways and Paths where Cattle dung and tread, and then clap it against a Wall till it be dry………).

For many commoners and other peasants, peat and turf were the main fuels and this can be evidenced from local histories and the like. For example, around Milnthorpe in north Lancashire, peat was the main fuel until *c.*1800. This is illustrated by an article of agreement made on 11 May 1687, between Mary Hinde of Auckenthwait, Spinster, and Thomas Cragge of Cragyea, in Auckenthwait, which concerned 'a dwelling house and garding'. This stated

> that Thomas should buy the property for £8 providing that Mary Hinde shall use occupie and possess ye Chimney and ye body of ye aforesaid house during her naturall lyfe … and that ye said Thomas Cragge shall give and lead cart to ye doore of ye said Mary Hindes house five cart full of peats or turfs in every yeare during her natural lyfe.
>
> *(Bingham, 1987)*

For ordinary people free use of peat fuel was important to their survival.

Fuel needs

Humans need fuel to live and as settlements grew from villages and farmsteads to towns and cities, as discussed in Rotherham (2005), these demands grew and became more complicated. For millennia, the challenge of keeping warm and having fuel with which to cook have been central to human survival. Only recently has the luxury of electricity down a cable, or gas or oil down a pipe, or even coal delivered to the door, been possible. In times past, in some cases even quite recently, many people went and got fuel from the local peat bog or turbary.

The importance of fuel to people has been discussed by Williamson & Warde (2013), and Warde & Williamson (2014). This was considered with regard to peat-fuel by Rotherham (2005, 2009, 2020), Today ideas of 'sustainability' consider the delicate balance between people and natural resources. However, in history if this failed and nature was exploited unsustainably the consequences could be devastating. In the Western Isles of Scotland for example, when peat fuel ran out and soils were exhausted people starved. In England, urban dwellers might turn to alternatives like coal but rural communities could not do so unless the supplies were close and also available to them. Until the 1500s, most mineral coals came from coastal outcrops and sea-borne transport, and so, for habitats away from ports it was prohibitively expensive. Rural populations used other, more readily available fuels, like peat, turf, and furze. These were supplemented by small wood and heather that were the main fuels of common people. However, in Britain and elsewhere in Europe, fuel was important and often scarce for long periods. From around the 1400s to the 1800s, the climate deteriorated, and in Britain, for instance, it was incredibly cold. Life was hard and without fuel, one died; turf was available free, but for the cutting and preparation though, you had to have the right to take it, and common rights were carefully guarded.

The peat or turf consumed by a single household for fuel, litter, and other purposes was around 8,000 turves per year. Six tons of peat was equivalent to one ton of coal, and peat stacks could be as big as the cottages. In East Anglia at Martham, a household used 5,000 to 8,000 turves per year, and at Scratby, 10,000 per year. Fuel was needed for both cooking and warmth and therefore was required all the year-round. Traditional peat-fires burnt all year and never went out – some reputedly burning continually for centuries. The peat-fire and the hearth were central to life in the uplands and for many in lowland areas too, especially around heaths and fens.

Apart from domestic peat-use, there was also commercial harvesting, the most dramatic English example being in the medieval Norfolk Broads. This landscape was long-considered 'natural', but in the 1950s abandoned medieval peat workings were identified cut out of the one-time huge raised mire. The peat was cut to supply Norwich and its Cathedral Priory with 400,000 turves per year during the 1300s and 1400s. Similar use occurred in other areas but may be less obvious. According to Arthur Young (1804), the fuel allotment was 1.75 acres at Northwold in Norfolk and produced 12,000 turves a year. This was the calculated consumption of one cottage hearth. However, common rights of turbary were jealously guarded and cottagers of twenty years standing but without rights were treated meanly. Controlled by the fen reeves they were awarded permission to cut only 800 turves each year which was nowhere near enough to keep a hearth alight. Fuel generally, and wood, in particular, could be very scarce.

According to Wright (1964), if wood was unavailable, then,

> Peat was the only alternative fuel to be had in any quantity. There is even more of it in Britain than in Ireland, despite the poets and the travel brochures; but it was too bulky to be carried far from its source. Peat burns readily; its merit is to smoulder without a blaze, though this makes for a smoky fire. The 'peat-reek' is pleasant at a distance, and a whiff of it is not unwelcome in one's Harris tweeds; but in a 'lum' cottage its pungency is dreadful. Peat has been 'coked' to destroy the reek, powdered, mixed with pitch or rosin, and compressed into bricks that were claimed to be better than coal. Lacking true peat, some burn turf, or parings of peaty soil with roots of heather and gorse. (Peat and turf, accidentally ignited, have caused slow, widespread and unquenchable fires, which have even consumed whole villages.) Dried cowdung is a good fuel, and the scent of its fire sweeter than might be supposed; it found some favour during the wood famine, but to burn dung that ought to enrich the soil is bad husbandry.

Concerned about woodland depletion and lack of fuel, Rev. William Harrison writing in Tudor times (in Withington, 1899) observed that

> Howbeit, thus much I dare affirm, that if woods go so fast to decay in the next hundred years of Grace as they have done and are like to do in this … it is to be feared that the fenny bote, broom, turf, gall, heath, furze, brakes, whins, ling, dies, hassocks, flags, straw, sedge, reed, rush, and also seascale, will be good merchandise even in the city of London, whereunto some of them even now have gotten ready passage, … Of coal-mines we have such plenty in the north and western parts of our island as may suffice for all the

> realm of England; and so they do hereafter indeed, if wood be not better cherished than it is at this present … their greatest trade beginneth now to grow from the forge into the kitchen and hall, as may appear already in most cities and towns that lie about the coast, where they have but little other fuel except it be turf and hassock.

Peat and turf were often 'free' but for the cost of cutting and transportation. In Scotland, and in Dorset too, they might be used as a valuable 'free' resource to help support vital and potentially expensive community services such as that of a schoolmaster. The writings of the schoolmaster Rev. Mr James Dingwall (*The Statistical Account of Scotland 1791–1799, Parish of Far (County of Sutherland)*) stated that "….He also has his peats cut, dried, and brought home free". Suffling (1885) described the value of peat,

> … this is turf, and its lower brother peat. This peat, or, as it is here called, 'hovers', is, when properly dried, a capital and economical substitute for coal. It gives off a blue smoke when burning, and this, as it rises from the cottars' chimneys, wafts a rather pleasant perfume in the air, which is a great improvement on the soot-laden, evil-smelling smoke of the metropolis. A peat-ground, properly managed, is a rather valuable holding, as may be gathered from the following statistics. The peat blocks, when cut are about 4in. square (shrinking by drying to about 3¼in.) by from 2ft. to 2½ft. long (the depth of the boggy surface soil). Each square foot, therefore, produces 9 'hovers', each yard 81, each rod 2450; and, consequently, each acre the enormous number of 392,000 hovers. As these are retailed at from 1s. to 1s. 6d. per 100, a good profit must be realised.
>
> *[Value = £196 to £294 per acre]*

In a classic study of the prices of commodities in England over a period from the 1300s onwards, Thorold Rogers (1902) examined peat and other fuels. He noted turf used in 1337, with amounts as the thousand (1,200) or the last (12,000), and there were varying descriptions and prices. In Cambridge in 1334 for instance, turf was specified as heather, and the average price was about 9d per thousand. Between 1415 and 1450, the Cambridge colleges used turves extensively bought by the thousand at around 2s. Turf was also frequently used in Southampton and Kent.

Fuelling towns, cities, and industry

It surprises many people that peat was not merely the fuel of the rural poor in Britain, but widely used in towns and cities also. In many cases, this ended so long ago that few records or memories remain, but there are obvious examples. Reasons for using peat included price and accessibility and that alternatives such as wood were often protected for other uses. However, there is a basic matter too of building technology and construction. Until the 1500s or maybe the 1600s, the poorer dwellings of most people did not possess chimneys and so burning mineral coal, even if available and affordable, was not an option.

Norwich with its ecclesiastical community was famously a centre that made substantial use of peat and turf from the Norfolk Broads. However, usage was far more widespread but this is now less recognised. Peat fuels were used extensively in, for example, Carlisle from the Solway Firth mosses, and in Cambridge from the Fens; York was fuelled from Askham Bog and other nearby turbaries such as Tillmire, a usage going back to Roman times. We have interesting medieval records of peat turf used in York in 1388, with peat or 'turbarum' brought into the city by barge to supplement York's turbary on the Tillmire near Heslington. The latter was running out of easily extracted peat and by the seventeenth century the volume of peat turf being consumed increased with small boats carrying peat from South Yorkshire's Thorne Moors (Rotherham, 2009, 2010).

Further evidence of commercial peat fuel activities comes from 1643, with reference to the supply of turf in York. Nine men were accused of "selling turfs contrary to my Lord Mayor's price". They admitted their guilt and were fined three-quarter pence each "... according to an Order made in the like case the 9th day of November 1593". The 1593 reference related to four men reprimanded and fined for a similar offence. In the 1643 entry for fines paid to the Corporation of York, it was stated that money was received ".... of watermen for selling turves before the price was set by my Lord Mayor 8/-".

The York fines were in November and perhaps a time when the cold was making turf especially important and a premium price could be charged even if illegally. Peat was worked across a wide area of the old West Riding of Yorkshire during the seventeenth century with references indicating 'Turf Mosses', to 'wayne leades of dryed peates', to the use of cartes for carrying dried peat, and to 'the turfpitt' from places such as Giggleswick, Thruscross, Barwick-in-Elmet, Bentham, and Fishlake. In the latter case, the notes refer to the theft in 1652 of "one Catch loade of Turves & wood" with a total value of £7. This was by four men one of whom was referred to as an 'Airmyn Waterman', and the 'Catch' was a ketch; this was one of the traditional open sailing trading barges of the Yorkshire rivers. There were even turbaries close to the centre of present-day Leeds. Along with domestic uses, peat and turf fuelled industries, mostly metal smelting. The suitability of peat fuel varied with the task and its industrial applications were often of mixed success. It was sometimes a back-up fuel should wood or charcoal not be available.

How peat was worked and the tools of the trade

To extract peat from a bog or fen, generally, the site must be relatively dry though there are some exceptions to this. The Dutch, for example, used dredges to take wet peat from the polders, and French tourbe cutters applied long-handled tools to extract peat from under-water. However, in Britain this wet extraction was localised with some cutting in Somerset being the scooping up of wet peat with a broad spade. Mostly however, water tables were permanently or temporarily lowered to facilitate cutting. This was generally easier on higher ground where a simple drain could be cut downslope to drain the excess water off. Individual bogs could easily be drained, at least in part, and the dry edges worked or cut for peat. In low-lying areas, as discussed earlier, this becomes more difficult because unless you can drain an entire landscape, your neighbours' water flows back onto your land. Furthermore, when cutting, even around the fringes of large raised mires, there is always the risk of destabilisation and a catastrophic peat-slide.

However, once the land is drier, then cutting, extracting, and drying peat or turf from fen or bog becomes possible. The peat is carefully dried and prepared and then eventually carried away for storage, sale, or use. There can be serious difficulties in the process, which is why many peat sites were un-reclaimed and unworked for many centuries. One issue is that if you drain a mire then it dries and shrinks and so your site becomes lower than the surrounding land and water flows back in to make it wet again. Active removal of water to a drain higher than the original ground requires pumps and as described earlier these developed from the 1500s wind-powered or animal-powered pumps to those of the nineteenth and twentieth centuries driven by steam, petrol, or diesel and ultimately by electricity.

In regions like the Somerset Levels and the Cambridgeshire Fens, largely from the 1700s to the 1900s, the now-familiar modern landscape was imposed on the extensive wildlands of fen, bog, marsh, moor, and water. The land was 'enclosed' by parliamentary acts with each administrative area (generally a parish), requiring its own legislation to enclose its commons. Within this process it was possible to award areas set aside as fuel-allotments and poor-lands for the upkeep of the landless poor (the incentive for landowners and the wealthier commoners was that this burden otherwise fell on the local parish to fund). Those with limited common rights might receive small pockets of enclosed land as part compensation for loss of privileges. In some areas, the process led to complicated systems of drains and ditches (in Somerset, for example, they were called 'rhynes') to divide the land into small rectangular units with occasional oddly shaped leftovers.

The transformation by enclosure generated 'modern' landscape patterns of roads and farms following a basic layout repeated across vast tracts of lowland countryside. The larger individual landowners claimed the lion's share but also took on the greater expenditure of draining and fencing. Smaller plots were charged in proportion but for some poorer commoners, the work was too much of a burden to undertake. Some of the enclosed parcels were peat bog or fen and still very wet with several metres of peat. However, whilst some of these areas were difficult or expensive to work, others presented entrepreneurial opportunities. Given the resource now in a drier and drying landscape, by the late 1800s, many peat sites in regions like Cambridgeshire and Somerset, were being worked both for domestic fuel and as commercial enterprises. This cutting was initially for peat fuel to use or sell, but then to fulfil the growing demand for animal litter. Gradually, though quite late in the day (mid-twentieth century) there was a growing

FIGURE 13.3 Hugely destructive industrial peat-cutting near Dublin in Ireland in the early 1980s.

market for horticultural peat. During the mid- to late-twentieth century, cutting moved from traditional hand-winning to commercial machine-milling, and then highly destructive open-casting. A consequence was that in many places long-standing traditions were lost almost overnight (Rotherham, 2005, 2009) (Figure 13.3).

Between early enclosures and the industrial, late twentieth century, there was a time of sophisticated, complex, hand-cutting of peat. The techniques and implements used had evolved over centuries, and probably originated from before the Roman period. However, it was the latter who invented the main implement for peat-cutting, that is, the metal spade. Wooden and bone peat-spades have been found in peat deposits in both the Cambridgeshire Fens and the Somerset Levels. However, the metal spade was invented for Roman military use and put to great effect by legionnaires to speedily dig defensive structures. The basic tool, modified and developed over many centuries, has many regional variants and often made by the local village blacksmith. However, across Britain and the European Atlantic coast the general techniques and tools were broadly similar (Jenkins, 1969; Gailey & Fenton, 1970; Fenton, 1986).

Cutting the peat

In preparing the peat grounds to be cut, the first step is drainage to make a site workable, reasonably dry, accessible, and safe. Safe access is needed on the peat-ground or turbary to get people, equipment, and often animals (for instance donkeys or ponies to carry the cut peat) to the place. Within the peat-fields there need to be working areas and places to rest and eat, as this was hard, time-consuming work. The working areas needed space to treat and prepare the cut turves or peats and stack them to dry. Whilst drying, they were turned, and then placed into larger stacks and were dried some more. Finally, the peat was taken off-site by sledge, cart, or pannier and then stored in covered stacks or barns close to the farm, cottage, or other suitable buildings if the peat was for re-sale. From very wet sites the cut peat might be carried out on special barges along drains functioning as canals. Peat was then either used by the cottager or farmer, sold at market, or else door-to-door.

The approaches used in terms of the tools, the style of cutting and preparation, and even the people involved varied considerably across the European peatlands. Even throughout the British Isles, there was great variation and local and distinctive styles and character to the peat-cutting. It is worth just considering a few specific examples.

The 'beckett' was the main peat-cutting tool in the English Southern Fens (Day, 1985). This was a slim straight spade rather like a cricket bat but with a metal flange set at right angles to the main blade. The tool made two cuts in one movement and was easy to push by hand through the soft peat. Across the Fens, tools of many different shapes and sizes were locally and individually made with willow shafts, and blades got often from cast-off iron tools and other local scraps. The blade had wood exposed to help turf slip off more easily, and the flange was around an inch narrower than the blade width. The resulting sharp-bladed tools cut tidy brick-shaped turves easy to handle, carry, and dry. With commercial cutting the dried turves were sold by measure. In the Fens this standard was based on a small beckett used at Isleham Fen in the 1850s, and this cut a standard turf measured fourteen inches by three inches. Small turves dried quickly but took longer to cut; at Wicken, the blade was about eighteen inches by four-and-a-half inches. After shrinkage, this cut a turf of eleven inches by three-and-a-half inches. Isleham small turves were sold in a thousand and Wicken large turves as six hundred 'to a thousand'.

An older fenland peat-tool was the 'moor spade' or 'sharp shovel' designed to be used with a strong shove from a foot. Some were made entirely from iron which meant the heavy weight made digging both hard and slow.

The process of peat-cutting was carefully implemented and with some variations from region to region. In the English Fens the digger made out the line of his cut and then used the spade to clear surface vegetation and material from the top fifteen inches or so. He then shaved the exposed peat to produce a level platform for cutting. A 'peat knife' or 'turf knife' was used to cut down through the sides of the pit or trench, or the cutter did this with the beckett. With two cuts for each turf the cutter worked backwards, one spit down (i.e. one depth of the beckett) and four spits across. As turves broke off they were easily lifted to the side of the trench which was cut the length of the field being worked. The cleared surface material was thrown back into the cut pit – carefully planned and thoroughly implemented, detail varied with the number of men working and the number of trenches cut at any one time. Carefully managed, a cut site could often be reworked thirty years or so after each cut. The most highly valued peat was the denser 'black peat' near the bottom of the cut. This commercial cutting differed from more widely practiced domestic peat-cutting which was very much a family- and community-based activity.

For commercial cutting, individual turves weighed about seven pounds, and the pit was a yard wide. The digger lifted to his side and left three turves height and broken ones on top. For commercial diggers paid by piecemeal, the speed, quality, and consistency were important. Things became more difficult as diggers moved to the second spit depth with a narrower trench (three spits wide). The work was physically demanding in wet peat with high humidity and clouds of biting insects. Piece-rate payment in 1900 was five shillings per thousand turves including opening, digging, turning, barrowing, stacking, and boating. Digging and laying peats were paid at two shillings and sixpence per thousand or a Wicken 600. Men worked from 6 am to 3 pm to produce about 3,000 Wicken turves per day with the record digging rate at 6,000 per day. Poor weather often slowed the job down. After digging and stacking, turves were left drying for about three weeks but varying with the weather. During this period peats were opened or dressed and restacked with gaps to allow air circulation which speeded the drying. Each turf was individually handled with a skilled digger dressing up to 3,600 per hour. Within six weeks, turves were sufficiently dry to barrow for stacking by the drain. Better drying weather was late summer and the priority for the early season was digging. Cut turves were tidied with the beckett or a trimmer, and any below standard called 'bits' were cast aside and sold cheaply to poorer local people. Turf was required all year round and supplying local households was important. To fill their carts with 1,500 turves a time, turf hawkers came to Wicken from Soham, Fordham, and Burwell. Their small carts were pulled by ponies or donkeys.

In Somerset, peat-cutters cut four-foot-wide 'heads' with lines marked with pegs (Sutherland & Nicolson, 1986; Anderson, 2006). The 'unridding' was thrown in the pit which was around twelve to eighteen inches wide, and cutters dug four layers or around three-foot-six-inches deep. They used various implements to mark out and dig and the peats were called 'mumps'. Flooding of the pits by water was a big problem in Somerset and early diggers ladled water ('lecking out') with a long-handled wooden scoop or shovel. Similar tools were used in early medieval times to shovel peat out from under-water. A widely used tool for taking off the top layer or skimming thinner peats was a long-handled 'turf spade' or 'paring iron'. In the Isle of Man, these were originally known as 'scraa-cutting spades' and were applied to cut thin turves for tiles or thatch. Following the cutting there were two processes called 'hiling' and 'ruckling'. The first was laying turf on the ground to build heaps of twelve turves known as 'hiles'. (On Crowle Moors on the Lincolnshire/Yorkshire border these were called 'walls'). The second process was stacking dry turves as 'ruckles' or 'cocks' in Somerset, or 'pyramids' at Crowle, and 'peat stacks' or 'ricks' elsewhere (Rotherham, 2009).

In Scotland, Ireland, and the Isles, as in the English Fens, peat-cutting was an ancient tradition going back to Viking times or before. In both countries cutting was associated with periods of land 'improvement' in the 1700s and 1800s and with starvation, a crisis in the rural economy, and problems of over-population. By the 1900s, larger-scale cutting of peat bogs was organised by government bodies such as the Land Commissions, Congested Districts Boards, or improving landlords. These bodies organised drainage and construction of access roads and tracks, and the layout of peat-faces or 'facebanks' that were to be worked. A main drain was opened across a bog, deepened year-by-year until the bog was 'bottomed' by the drain. Sub-drains were cut parallel to the main drain and cross-connected to it. In working a raised bog, the cutters described four different and distinctive zones or layers. The top of the profile was the 'clearing' or 'top scraw', below which was 'white turf' of undecomposed material, and then 'brown turf'. At the very bottom of the mire was rich black fen peat.

The tools of the Irish peat-cutters included a special implement called a 'flachter', 'skroghoge', or 'straw cutter' (Evans, 1942). This was used in the north of Ireland, imported from Scotland, to cut away the heathery 'scraw' (from the Irish *scraith*, a green sod) from the bog surface before the actual cutting. As in the Fens, these were very individual tools made by local blacksmiths with handles of pliable wood and metal blades from scrap iron. The flachter was essentially a modified breast-plough and though not widely used in Ireland the process was sometimes important with up to two metres of bog surface cut away before turf-cutting began. It was used particularly in the mountain bogs of north-east Ireland. In carefully managed sites, the scraw was reserved and placed back on the bog after cutting. The Irish, like the Highlanders, had specialised vocabularies related to bogs and peat-cutting, with many regional variants. A turf-bank in Scottish Gaelic is *bac móna, bakki* being Old Norse for bank; the Scottish Gaelic for turf spade is *tréigeir* from the Old Norse torf-sceir or turf-cutter. There are Norse influences through the Viking connections with Ireland directly and via the Western Isles.

In Ireland most turf was cut with a turf spade or slane (*sleaghán*), similar to the Fenland beckett (Feehan & O' Donovan, 1996). After clearing the top scraw, the slanesman could begin cutting. The winged slane blade allowed a double cut with turf extracted in one swift action. Slanes were made locally by blacksmiths or in bulk in spade mills, with two main types. The breast-slane winged or not was used extensively in blanket bogs with the cutter facing the turf-bank and required reasonably dry conditions as cutting moved along. The turves were cut with a horizontal stroke into the bank and then vertically by a second cutter standing on top of the peat-bank. Cut turves were placed on top of the uncut peat with around sixteen 'boxes' of fourteen by six by six-inch turves per day. The foot-slane produced turves of ten by five by five inches and were used for deeper valley-bottom bogs. The sods were cut on two sides by vertical strokes of a winged slane, the bottom breaking as the blade lifted. The sod was thrown to the waiting barrow-man – quick and simple. The barrow-men took peats away to spread on the dry cutaway area or 'low bank', in front of the turf breast or 'high bank'.

In Scotland and the Islands, cutting was with a 'peat iron' or 'flauchter spade' made by a local blacksmith and similar to Fenland becketts or Irish slanes. According to Alexander Fenton this was noted in 1493, but the tool is even older (Jenkins, 1965; Fenton, 1986). Peat-cutting for fuel and to fertilise the Highlands and Islands infields goes back further still. The tools are locally distinct, so in Shetland it was a slim, straight, shafted spade called a 'tusker' with no foot-peg and operated by the arms. Orkney and Caithness were similar to Shetland but sturdier and operated differently. A description from Caithness in 1812 indicated two different ways of cutting peat, horizontal if peat was one- to two-feet thick, and perpendicular if deeper. The tusker or 'turskill' had a wooden peg to allow use of foot-power. The cutter held the handle with both hands and pushed the foot to drive the spade into the moss. A woman stood at the base of the peat-bank to take the peat as the cutter jerked it away from the bottom. She grasped the turf and threw it to the right side of the bank. Here others, perhaps children, spread the peats to dry. Such good quality black peats were called '*tusker* peats' and were much-prized items of sale; poorer quality turf was for home consumption. In Orkney, when poorer quality hill-peats were cut, a modified tool like a tusker was used. This had a broader blade with a vertical wing to one side, and was called a 'luggie', meaning the 'eared spade'. Across the Highlands and Islands was a great array of peat-spades some similar to basic tuskers, others with distinctive wings and flanges.

In north-east Scotland, spades often had broad blades with a round mouth, any wings were short. Two types of spade are described, the 'breist spade' or 'breasting spade' and the 'stamp spade' or 'underfoot spade'. The former was used horizontally by cutting into the peat-bank face, the cutter standing below. The latter was handled vertically and stamped down with the foot from above the bank. Most cutters had both types for different situations. In central Scotland, the pattern changed again with square-mouthed wooden blades shoed with iron cutting edges. Further into south-west and south-east Scotland, were local variants, though evidence of the latter is sparse as peat-cutting died out long ago. By the late 1800s and 1900s, factory-made peat spades were manufactured in Scotland or imported from England, often reflecting local variants and demand.

The approaches and techniques applied to turf cutting throughout the British Isles were a mix of those applied in lowland fens, or bogs and uplands of the north and west (Fitzrandolph & Hay, 1926). There were many local variants and distinctive local approaches, too variable to describe, and over time they have mingled and altered. In recent decades tools and implements have moved widely as people have travelled. Across Europe and among Britain, continental Europe, and North America, there were exchanges of ideas, technologies, approaches, and the tools applied. The tools and styles have been taken around Europe and across the Atlantic.

Peat-cutting in the annual cycle

Traditional rural communities frequently organised the tasks of living and farming following an annual cycle based on seasons, weather, crops, activities, and labour availability for specific tasks (Knox, 1985). In temperate peatlands, cutting of peat and turf was a key part of the cycle carefully embedded into other activities. Labour-intensive work happened when people were free to do it. Most site work was from May to August, but varied with local climate, for example, earlier on the English Dorset heaths, and later in the Scottish Cairngorms.

On peat bog or fen, there were key stages of work, with checking and preparing peat-grounds early, so when weather and labour permitted, cutting could begin. Peat was cut, prepared, and stacked to dry and then drying turves were carefully and regularly turned to be dried thoroughly. The latter was work for women and children, and essential for next winter's fuel supply. Finally, dried peats were carted for storage and use.

Why peat-cutting ended

Mostly, turbaries were abandoned because of exhaustion or as cheaper, easier fuels became available. In mid-Wales around Rhayadder, for example, valley-bottom bogs were exhausted long before mineral coal arrived by train in the 1860s/1870s. Hilltop turbaries were exploited and these were abandoned when coal became available. Archaeological heritage of peat-cut sites remains in the landscape as evidence of usage and abandonment in the mid-1800s. In the Yorkshire Dales, and the North York Moors extensive peat-cutting finished around the 1930s or 1940s (Hartley & Ingilby, 1968, 1972).

Peat fuel use was often not sustainable in the longer-term and when peat ran out alternatives were found or communities moved on. Changes were often triggered by competition from other cheaper, more accessible, or better-quality fuels. Mann (1993) in *The First Statistical Account of the Scotland for the Parishes of Anworth and Girthon, 1792*, noted that in coastal areas of south-west Scotland,

> Peats, the fuel used by the farmers and cottars, are dear, owing to the distance of the mosses, and, the bad roads which lead to them.........Coals are the general fuel here. They are imported from Whitehaven, Newcastle, etc. and run from 30s to 40s a ton.

Smith (1951) stated that in north-east Scotland,

> Peats are not used so widely as in former years. Prices are high normally £2 and over per cartload. Coal is brought to the door by motor lorry from Aberdeen and sells at the high price of £5 a ton for English coal and £4–15 for Scottish coal.

Accessibility to fuel transport was the key, and locations difficult for transport or remote from the coastal distribution or local coal supplies maintained peat-use for longer.

In some cases, the industries needing fuel closed or changed. So, loss of demand from industries like lead-smelting in the Yorkshire Dales (Raistrick & Jennings, 1965) or rural depopulation affecting domestic usage, were other causes. Exhaustion of supplies, the hard work of extracting and preparing, and financial considerations (especially cheap and easily accessed energy supplies from coal, electricity, oil, and gas) were all important in people moving away from peat. Indeed, the survival of peat-cutting to the mid-1900s was often because fuel was cheap or free (Hartley & Ingilby, 1968, 1972). In some cases, manufactured products from peat fuel were lower grade and lower value than those from competing fuels.

The enclosures and fuel allotments

The enclosures of English commons, especially the parliamentary enclosures of the eighteenth and nineteenth centuries, brought major agricultural improvements, but at a cost. The effects on the poor and needy were dramatic and when commons were enclosed and traditional rights lost, there were attempts to compensate the cottagers and the landless poor. This was done often by means of small plots of land held in trust for the benefit of the poor as charitable 'poor allotments'. However, it was important to note that allotments were often a commuted right granted as compensation for the loss of common rather than a gift of charity.

Around the settlements these allotments were often called Town Lands, Poor Lands, or Fuel Allotments. Common rights of the medieval manorial system were in two broad categories, those 'appendant' (related to land ownership) and those 'appurtenant' (related to ancient cottages). These gave rights of common on the village 'waste' including pasturage, piscary (fishing), estover and firebote (wood gathering for house repair and for domestic fuel), and turbary (cutting of peat and turf). Other common rights covered cutting of furze for fuel and mineral or stone extraction. There were additional rights granted of 'common in gross' vested in individuals through a grant or by prescription (twenty years unchallenged use). For fuels, there are key factors to be considered like types of fuels available and accessible, calorific value and quality (i.e. how much heat they produce and for how long), and costs in effort and energy expended to gather and process. Wood, gorse or furze, fern or bracken, ling or heather, and even cow dung were all collected along with peat or turf. The cost of buying fuel rather than the effort of gathering it was important and variable. In Victorian Berkshire for example, it was estimated a family could cut a year's fuelwood in one week but to purchase the same would take a tenth of a labourer's annual income (Davies, 1785).

Coal might be cheaper but depending on quality and the type of fireplace, it was not always favoured. Simple hearths even with a chimney, (which most but not all cottages had by the 1700 and 1800s), did not burn coal effectively and generated noxious fumes. Only with metal grates and tall narrow chimneys was coal the better fuel, despite it having around twice the calories of wood. But wood was a cleaner fuel to handle and burn and had the advantage that its ashes could be sold to industry or agriculture. However, wood was often protected for other purposes within a manor and in the age of charcoal manufacture for iron-smelting, it was in limited supply for domestic use. Peat and turf burn long and slow, but produce the traditional 'peat reek'. Furze was commonly used by the poor and was similar to wood but quicker, hotter, and short-lived (Rotherham, 2005, 2009).

Examples of fuel allotments

The Graveship of Holme, West Yorkshire

This is almost England's last community-organised peat-cut, with others in North Yorkshire and Cumbria already being quashed. This was discussed in Handley & Rotherham (2011). The Graveship, established at the time of Henry V, is overseen by the Constable and his assistant and each year those owning a hearth in Holme Township can register rights to cut peat and a small area of turbary is allotted. Grazing rights are controlled by the constable of the Graveship, and fuel rights are for individual domestic use not commercial exploitation. In times past, rental from the allotments was given to the parish poor – £20 in 1887. By the late 1990s, the rent of £300 per year was available for road maintenance. Around a hundred people still registered their rights in the late 1990s, with only three (Isle of Skye, Hades Peat Pit, and Harden Peat Pits) of the original five turbaries still in use, around twenty-two hectares. Rights are claimed at the end of Easter with cutting in May and June. In recent years the practice may have ceased. Mass 'wastage' of peat across one flank of Holme Moss was wrongly attributed to the peat turbaries. In fact, this was caused by two former soldiers discharged after World War Two who set up a business on the moss to supply peat fuel to factories across Yorkshire during the post-war energy crisis.

The North Devon Turbaries

A remarkable example of a poor-lands turbary is that of the Stockland Turbaries in North Devon near Honiton maintained by a local project as nature conservation areas (Anon., undated). This is described by Rotherham (2005, 2013a). The Turbaries were established by an Act of Parliament when local commons were enclosed in the early 1800s. The map accompanying the award was dated 1824 and the conditions of the Turbaries were described in *Reports made to the Charity Commissioners* included in the *Return to an Order of the Honourable The House of Commons* 26 July 1905, with Reports on the Parishes of Stockland and Dalwood dated 19 July 1834. Lands and other properties generated income and this could be used for ".....any other pious and charitable use or uses within the said parish as the feoffees for the time being should think fit". These included the Stockton Ecclesiastical Charity, the Stockland Educational Foundation, and the Stockton Parish Lands (General Charity); collection and distribution were overseen by appointed trustees.

The total annual income in 1908 was just over £44, of which £4 went to the Parish coal fund, given away at Christmas. In addition to the coal fund there is mention of turbary lands.

> By the Inclosure Award already mentioned there were allotted and awarded to the churchwardens and overseers of the poor of the parish of Stockland for the time being, and their successors for ever the twelve several parcels of land hereinafter described (that is to say), ...

The document then goes on to give details of individual turbaries: Bucehays Turbary, Quantock Turbary, Short Moor Turbary, Huntshays Turbary, Hamer Hill Turbary, and Shore Turbary.

The memory of their use as working turbaries, abandoned some time from 1910 to 1930, has long disappeared, but a photograph of turf-cutting in 1914 shows this at the end of its usage. Loss of rural labour in the war effort, the Depression, improved road and rail transport, and the availability of coal and coal-burning stoves meant abandonment.

Industrial peat-cutting, extraction and exploitation

Since the beginnings of early industry, peat and turf have been used in many different ways in industries (e.g. Collingwood, 1912; Blizard, 1917; Donald, 1955; and Greeves, 1981). These include obvious uses in agriculture but also in metal smelting, and with distillation treatments to extract valuable organic chemicals. This was especially important in pre-mineral oil economies. From a very early time it was realised that peat – like wood, and like coal and oil later on, could be distilled to produce organic chemicals (Kerr, 1905).

Peat was used in primary industries such as the smelting and extraction of metals, and in the subsequent factoring of tools and objects. Fenland blacksmiths used peat rather than wood, mineral coal, or charcoal, and in Wicken in the 1930s, Bill Redit was the last to use turf burnt with wood. He built fires outside close to the 'tyring platform'; the hot metal 'tyre' swiftly drawn from the fire hammered onto the wooden wheel, tightened by cold water shrink-fitting metal to wood (Day, 1985).

In Ireland there were many examples of industrial peat use from fuel to chemical extraction. For example, between 1906 and 1922, using 8,000 tonnes of hand-cut peat per year, the Marconi Trans-Atlantic Wireless Station at Clifden was powered by turf-fuelled steam engines. Peat gas was another product and, for example, powered the Robb Linen Factory at Portadown, taking 3,500 tonnes of peat each year.

Chemical extraction

Peat is a potential source of complex organic compounds and, therefore, much sought after by early industry before petrochemicals became widely available.

There was considerable interest in processes and industrial works to distil peat in ways similar to charcoaling wood, to produce volatile oil-like products. Industrialists in the pre-petrochemical era saw peat as a source of volatile organic chemicals such as tars and paraffins. There was already an extensive industry to distil chemicals from wood in processes linked to charcoal manufacture. So, the vast reservoir of peat seemed an obvious target for experimental industrialists.

Throughout the 1800s, in the extensive bogs of Germany and Poland chemical engineers sought novel ways to exploit this potential. The vast peat resources of Russia and the former Soviet Union were also exploited on a large scale. Alongside peat-mining for fuel there were many other applications.

The peat was used to make dyes and tannins, for candles and fertiliser, and for alcohol and ammonia. These processes were copied in Britain with peat chemical works on Dartmoor (producing 'naptha' and gas), on Sheffield's Ringinglow Bog in the Peak District, around Thorne Moors in South Yorkshire, at Fenn's Moss in 1923 by the Bettisford Trust Company (taking black peat "....to extract paraffinoid and other chemical products") and in the Scottish Islands. None of these lasted very long, the processes being costly and inefficient, and of course hugely destructive to the peat bogs. Many companies lodged patents for specific processes, for example, the Bettisford Trust Company at Fenn's Moss in 1924 patented a process of 'destructive distillation'. This used high temperatures to generate benzenes and lower temperatures for paraffin. The Lewis Chemical Works producing peat-oil in the mid-1800s was a particularly famous example from the Scottish Isles.

By the late 1800s and early 1900s, carbonising peat processes were developed to extract chemicals. These included Wieland's carboniser in Oldenburg which, with three ovens, produced ten tons of peat-charcoal every day. The peat was broken down into the following constituents: turf charcoal 30%; tar 2.5%; tar water 32.5%; gas and losses 35%. From these products other chemicals could be manufactured with the tar fraction producing lighting oil, creosote, paraffin wax, pitch, methyl alcohol, and calcium acetate. Ammonia and acetic acid were also produced. In the years before effective extraction of coal tar from mineral coal this destructive distillation of peat was very exciting.

Peat-gas was also produced in Europe to provide heat and light, though it was not commercially viable as an enterprise. However, it was used in Ireland to light several small towns such as Mullingar in the 1850s, and again during the emergency efforts of World War Two (Figure 13.4).

Peat charcoal

There are problems with using peat fuel in that it produces little energy for a large volume. It is also bulky and friable and therefore difficult to transport from sites of production to its place of use or sale. Turning peat into peat charcoal both reduced volume and increased its thermal capacity. Charred peat, that is peat combusted under conditions of carefully restricted oxygen availability, is

FIGURE 13.4 Industrial peat-mining increased dramatically in the mid-twentieth century as seen here in Scandinavia 1941.

particularly suitable for metallurgical processes and also for the production of hydrogen disulphide and activated carbon. The results however, were sometimes unsatisfactory (see Peter, 1994 for example). It has a high reaction potential due to its great pore surface area, low sulphur content, and low ash content. Activated carbon made from peat has more than a thousand square metres of surface area per gram. It also has the advantage of being very light with 400 litres weighing only 100 kilograms.

From the 1600s onwards, scientists across Europe experimented with ways to char peat to make useful products. With severe shortages of wood-charcoal for metal-smelting, there were debates about the potential to use peat-based charcoal. However, there were severe problems of achieving an even burn and the friability of peat-charcoal meant it blew away easily in processes where updraft was important. However, carefully prepared briquettes of peat charcoal proved superior to wood charcoal. There were great successes but also failures.

In some cases, the charcoal was produced for a specific purpose such as for iron-smelting at Creevelea in Ireland in the mid-1800s, where there were bonuses of other by-products. In this case, the cost of producing peat-charcoal for smelting was more than covered by the value of peat petroleum and tar also manufactured. Peat-charcoal was produced in England, especially on the moors of the south-west, probably linked to local metal-smelting and a shortage of easily available wood-charcoal.

Other uses of peat

There were many and varied uses of peat and its products, and mostly these have been long-since forgotten. Some of these uses are introduced below.

Peat baths and moor baths

Many people are aware of peat used as fuel, and for horticulture, but other uses are hardly known. They are amazingly diverse and interesting. Furthermore, peat has always been a special resource and often associated with almost magical and mystical powers. Peat bogs have the power to preserve and embalm bodies and other organic items accidentally lost or deliberately placed in them. From early times they were viewed with horror, fascination, and awe, and to the present-day peat has a reputation for healing properties; some of these are well-founded. Peat and peat products are widely used in European health spas and from the mid-1800s to the mid-1900s was being promoted as a healing and therapeutic treatment. In Britain this was limited to baths at Buxton and Harrogate in England, Llandrindod Wells in mid-Wales, and Strathpeffer in Scotland (Rotherham, 2009, 2014b).

With diseases like consumption, typhus, and cholera taking huge tolls in industrial towns of the eighteenth and nineteenth centuries, people were desperate for health cures. From hot mud, hot and radioactive peat, and high-pressure cold-water jets, customers endured much in the search cures. Indeed, bathing in the complex organic chemicals of hot peat was good for the body and perhaps the soul.

By the later 1800s, straightforward water spa treatments were becoming unfashionable but there were new technical innovations which became popular. Electrotherapy involved patients having electric currents passed through them whilst in a bath or hot water, or in some cases if required, hot peat or mud. Treatment might be with rectal or vaginal douche and treatments were also good for rheumatism, arthritis, and gastric complaints. Electrotherapy cures like galvanism, the passing of electric currents through the patient's body administered by an attendant cum electrician, were offered. Attendants first checked the dosage on themselves to ensure it was at the right level; a rather haphazard method that sometimes ended with fatal results. This treatment supposedly toned the body and the nervous system. Faradisation with electricity had several variants but the basic approach was for the patient to sit with their feet against a large electrode. The other end of the circuit was attached to different target parts of the body in turn. In Germany, peat rich in radon was especially highly regarded, and along with radioactive peat, you could opt for the rectal or vaginal hot-peat douche whilst being electrocuted at the same time. Today you can purchase peat creams and packs to apply externally and to rub into skin affected by various ailments. It is suggested that the health-giving properties of peats derive from their rich chemistry and biochemistry with trace elements and complex organic compounds. These were exploited in products such as 'sphagnol' soap with special curative effects helpful for skin complaints. The extracted minerals are still marketed both in North America and in Germany as 'Mineral Life' liquid at $55 (Rotherham, 2009, 2014b).

Peat paper and card

Peat was used for making paper and card for various uses. Postcards from Ireland were printed with typical Irish folk scenes onto paper manufactured from the bogs themselves. Peat paper was made in Ireland from around 1835. By 1903, there was a factory in Cellbridge County Kildare that manufactured brown wrapping paper and card from peat-moss fibre and waste-paper pulp. In 1905, the Callendar Paper Company was producing ten tons of peat-paper a week, but this was not economical and so could not compete with wood or straw paper. Postcards from Irish peat-paper were popular around 1900 and were being sent across the world as a link to home for expatriate Irish, or a romantic touch of 'Irish-ness' for tourists. They were also especially popular at Christmas as touching reminders of home for Irish people now thousands of miles from the Emerald Isle (Rotherham, 2009).

Peat in construction

In Victorian Europe, there were experiments in making peat building blocks for houses using compressed peat and slaked lime, and it was also used for under-floor and roof insulation in the Alps and countries such as France for example until well into the twentieth century. Peat-and-tar roofing and pavement materials were also used.

Turf or sod was often used for constructing cottages for the poor because they were cheap, accessible, and available, and because its unique properties made it suitable for rough dwellings in cold and wet conditions. Across parts of Europe as well as in North America, 'sod houses' or 'sod cabins' with walls made from sods along with roofing, were widely used in some regions.

Turf, especially heather turf with the roots intact, was often used to roof poorer houses since these materials were free and easy to maintain, dealt with rainfall very effectively, and were very warm. Workers in other environments such as charcoal burners in English coppice woods made wigwam shelters covered with cut turf (rather than peat). Fenland cottages of timber and wattle and daub used turves with reeds for insulation; one such buidling in Wicken, demolished in the 1950s, still had perfect solid turves in its walls. Homes of wet peat or sod were common in more remote or poorer regions especially Ireland, made mostly with cut turves but sometimes with walls of standing uncut peat. However, this was most famously the case in the peat-houses of Flanders Moss near Stirling in the 1700s, when the shelters were built by the 'Moss Lairds' hired to drain and clear the peat-moss.

The evidence of such buildings was transient as were the dwellings themselves. They were prone to sinking back into the bog from which they were formed and were also, for obvious reasons, prone to fire damage. Some of the buildings were partly subterranean; others built of turf stacked on the bog surface. Similar constructions were also common in parts of Ireland, and remarkably some in Leck near Ballymoney, Ireland, were still occupied in the 1950s.

Peat and agriculture

Peat was used extensively for agricultural purposes with vast quantities of peat and shallow turf heaped and burnt to produce peat-ash fertilizer. This was sometimes part of the process of 'land improvement' undertaken as 'paring and burning'. In other cases, such as the lowland valleys around Reading in England, peat was extracted from nearby fens, taken to specific 'pits' and burnt to produce ash.

One of the main uses in the late 1800s and early 1900s was litter for animal bedding. With huge numbers of animals powering farms and a big demand in towns and cities for material to keep things clean, peat was ideal for this purpose. Once soiled, its nutritive qualities enhanced, it went on the land as fertiliser. At Thorne Moors in South Yorkshire for example, the English Moss Litter company extracted peat-moss up to the 1960s, and from 1923 to 1962, the Midland Litter Company took moss from Fenn's Moss near Wrexham. Historically, raw peat was widely used as litter by farmers and peasants, but wider usage started in the early 1900s. Processed and packaged as a commercial product, during World War One, with its absorbent, fibrous properties combined with antiseptic properties, it was harvested extensively as horse-bedding for military usage.

In this context, peat was also used as animal feed, mixed with green fodder and perhaps molasses. Again, its antibacterial properties may have had therapeutic effects and it was used either coarse or as a powder mixed into a cattle-cake.

Peat for horticulture

Today, horticultural peat usage is one of the most widely known demands for the resource. However, this was a relatively recent development which is linked to the massive growth in twentieth-century leisure gardening on the one hand, and the emergence of commercial horticulture and garden centres to service that emerging sector on the other. Nevertheless, from the 1930s, demand for horticultural peat supplied by hand-cutting grew rapidly. Then, by the 1950s and 1960s, there was increasing interest in commercial industrial exploitation by machine. This raised operations to another level and sites had to be drier to support big machines and narrow-gauge railways to extract bulk peat. Such scaled-up extraction required big drains and huge pumps and meant landscapes desiccated and ecology devastated.

In Ireland, not only was there a demand for horticultural peat, but this coincided with a massive expansion in industrial extraction of peat to fuel power stations across the country. Sites which had evolved slowly over the millennia with complex interactions between people and nature were destroyed overnight to leave barren, blackened landscapes.

Whisky and whiskey

One of the most globally famous uses of peat is to fire the kilns in whisky distillation and especially in the process of drying the barley before distillation. This imbues the liquor with a uniquely smoky peat flavour characteristic of the West Coast malts such as the Islays. Peat was also associated with illicit distilling in the Scottish Highlands, in North Wales, in Ireland, and even in England such as in South Yorkshire. Sustaining long-term use and exploitation of peat in malt whisky presents major challenges to the industry. Whilst distillers are probably aware of some of the issues of sustainability, this is a problem that will necessitate serious consideration in the near future.

Peatlands and people

It will be apparent from this short account, that over the millennia peat has had a fascinating history and that peatlands are deeply eco-cultural places. Also, in recent centuries, peat and peatlands have been hugely important resources for communities and even industries across much of Europe, and Britain in particular, but also around the world such as in North America. However, as peat is cut and mined at rates which far exceed any hope of regrowth, the utilisation of peatlands is ultimately threatened by the extraction and by the actual use of peat itself.

Given the long time to generate a peat bog, the types of exploitation witnessed over recent centuries are clearly not sustainable. Furthermore, the utilisation across the British Isles and the rest of Northern Europe during the late twentieth century was devastatingly destructive. Worse still, these impacts are mirrored across the world. Over the millennia skills, techniques, and implements used in the extraction and processing of peat were exchanged across Europe. Indeed, in Holland and Flanders, in Germany and Poland, in France, and Scandinavia, peat was used extensively. From Scandinavia the Vikings exported peat-cutting tools around north-western Europe, and the Germans later pioneered industrial chemical extraction. Peatlands and peat are truly international resources and their destruction has been an international affair.

However, the utilisation, whilst destructive, has left a legacy of implements, culture, landscapes, and language related to centuries of exploitation. This is itself now a threatened part of our cultural heritage. The history and culture of peatlands and their use was spread across the world by emigrating Irish, Scots, Welsh, and English, and by Europeans especially from the Low Countries. Peat-cutting tools, associated landscapes, and folklore are to be found in North America especially in the eastern part of the USA and in Canada too. But they are also scattered as far as the Falklands in the South Atlantic and other remote corners of the planet. In Britain today though, these are often forgotten times and neglected memories.

Peat itself is a fascinating substance and has been an important resource. This account focuses on the peats and peatlands of the British Isles and Europe specifically, and these are typical of the north temperate environment. Yet peats are formed and occur wherever plant materials collect in wetlands and cannot decompose, and they are found widely in suitable tropical and sub-tropical climates; even there, they have been badly damaged.

So, whilst this particular account includes a celebration of peat's importance, and the heritage and culture of peat-use, there are serious problems to be reckoned with. Remarkably, these are related to issues right at the core of modern society. The main concern is the important role of these peat landscapes in environmental processes such as water management and the release of carbon dioxide as a causative agent of climate change. It is clear that huge amounts of peat have been removed over the last 200 years in particular, and this has released colossal volumes of carbon dioxide into the atmosphere. Furthermore, in removing the peat that once great sponge, that both held back floodwaters and alleviated droughts, has gone. Finally, in a climate of energy shortage there is a renewed interest in peat fuel and its traditions. Hopefully, any future use might seek to be sustainable. Small-scale community usage may be carefully managed as a cultural heritage tradition, but anything larger is more problematic. Peat-mining is totally unsustainable and hugely destructive (Figure 13.5).

FIGURE 13.5 Vehicle damage to peatland on Big Moor, in the English Peak District and ironically a bog restoration project causing accidental damage to heritage.

PART 5

Peatland recovery and repair

14
CONSERVATION AND RESTORATION

> Restoration of peatlands is a low hanging fruit, and among the most cost-effective options for mitigating climate change.
>
> Achim Steiner UN Under-Secretary General and Executive Director
> UN Environment Programme (UNEP)

Peatland loss

Given the catastrophic destruction of peatlands, especially during the twentieth century, the consequences for the ecosystem service delivery have been severe and the outputs badly compromised. Belated recognition of the scale of the problem and the ongoing and incipient climate change has triggered considerable interest in restoring or reconstructing peatlands (e.g. Wheeler & Shaw, 1995; Parkyn et al., 1997; Cris et al., 2011, 2014). In England, for example, this has led to major projects in the lowlands such as the re-wetting of the Humberhead Levels, the Great Fen Project, the Lakenheath Fen Project, the Wicken Fen Project (Rotherham, 2013b), and Cumbria BogLife around Morecambe Bay and the Solway Firth (Rotherham, 2020).

In the English upland areas, there are numerous initiatives taking place such as Moors for the Future in the South Pennines, and the Yorkshire Peat Partnership further north. In the north-west of England, the Cumbria BogLife project has made great strides towards beginning the slow journey of long-term site recovery of upland peatlands. In parts of Germany and elsewhere in Europe also, there are extensive restoration programmes now underway.

Not only are these projects attempting restoration on previously unprecedented scales, but they are also developing our knowledge of the science and process of restoration and reconstruction. Emerging from coordination with the IUCN Peatlands Programme is an effective toolkit approach to reconstruction (e.g. Bain et al., 2011; Cris et al., 2011, 2014). Excitingly too, some of these projects represent examples of where the restoration of effective ecosystem services is being funded in part by commercial operators. Indeed, peatlands are among the few systems where the economic and other benefits to commercial stakeholders are unequivocally apparent. With the above in mind, the current European situation is perhaps one of greater optimism than for perhaps fifty years or more. However, this does not mean the problems have gone away, but more that we understand better the scale of the challenge and we now have the tools and some corporate willingness to address them.

The IUCN Peatland Programme '*UK Peatland Restoration – demonstrating success*' was produced in 2012 and provides a comprehensive overview of peat-restoration projects in the UK at that time. The parallel volume '*Global Peatland Restoration – demonstrating success*' was published in 2011 and gives a similarly structured review of worldwide projects (Figure 14.1).

In the wider landscape of both uplands and lowlands, that is, outside of restoration project areas, the problems remain and maybe getting worse. In former lowland fens, now intensively farmed, carbon is still haemorrhaging into the atmosphere on a colossal scale to add to greenhouse gas levels and cause degradation and loss of soils. It is now estimated that many of our most productive agricultural lands have only around fifty years of cultivation left in them. In upland zones too, many formerly wet moors, fens, and mires long-since drained,

FIGURE 14.1 Grip blocking in south Cumbria, England is one easy and effective way to begin re-wetting the bog.

have not been restored. These continue shed floodwaters downstream, to release carbon into the atmosphere, and to trigger brown discolouration in drinking water supplies. In many cases, it is not even realised that these sites were peatlands.

North American peatlands in the USA and Canada also suffered massive losses following European settlement in the 1700s and 1800s. However, again there are now major restoration projects in at least some of the damaged areas. Nevertheless, there is still an issue of considerable losses, for example, along the American eastern seaboard. Many of these were sufficiently early and so total that their former existence is hardly recognised.

There are various drivers for restoration including the recovery of biodiversity and ecosystem services such as carbon sequestration and water management. However, emerging economic drivers such as long-term tourism benefits, sporting interests including shooting, and sphagnum farming or paludiculture for horticultural produce, all present opportunities.

Often missing in many restoration projects (which focus on hydrology and ecology), is the safeguarding and conservation of the unique heritage of the peatscapes – the human footprint over thousands of years. Some of the heritage is in the physical landscape and the associated monuments, and some in the 'intangible' cultural heritage; these combine to generate a 'sense of place' which may help trigger tourism benefits. Again, in this context, frequently overlooked is the upstanding archaeology associated with past peat extraction industries. This is is treated with scant regard by those involved in the restoration. The ecological and landscape footprint of human peatland use over centuries is frequently etched into the modern mire, bog, or fen, but this remarkable evidence of human history is easily removed or compromised by restoration projects.

This chapter closes by raising some issues to challenge many, if not all, current restoration projects and their managers. How can we restore ecosystem services and at the same time recognise and conserve the cultural heritage and the archaeological monuments? One of the few upland accounts to address this and to help guide work on the ground is Bray (2015). For lowland sites the account by Berry et al. (1996) is another example of good practice although how much heritage conservation work has been undertaken on the ground is unclear. The various volumes of the North West Wetlands Survey, and the archaeological surveys of the Humberhead wetlands, also demonstrate the potential of such peatland areas.

Lost peatlands in the twenty-first century

The lowland fens

The story of the lost and damaged peatlands is one of human attrition over many centuries with the virtual eradication of nature over entire landscapes. In the English Fens, for example, it is the removal of several thousand square miles of wetland and a unique ecology. Furthermore, this eradiation applied to the wider landscape as well and so the total loss is beyond comprehension. None of us alive today has ever seen such a landscape, and so, in the words of David Lowenthal (Lowenthal, 1985), with 'the past being a foreign country', we can hardly imagine it.

In '*Lost Fens*' (2013), I wrote about just those fenlands in the east of England but noted the wider national picture to include wetlands in Somerset, in the Severn Valley, around Merseyside, Cheshire, Lancashire, and more. Additionally, it seems that the basic plot is always the same but each story has its unique twists. The demise of England's fens was mirrored across Europe, and through European settlement, it was exported globally around the world. What is described has been a global catastrophe.

These landscapes were at once dark, dangerous, unpredictable black waters, but at the same time productive, rich, and provided a sanctuary from oppression or persecution. The fens were, and indeed still are, contested spaces. Different stakeholders vie and compete with each other and with Nature for possession of land and water. In the last 200 years, a major change has been the move from local ownership of and subsistence on the resource to capital-based industrial exploitation. This transformation has involved a remarkable piece of political and legal manoeuvring which was at the heart of William Bunting's objections in the Northern Fenland. In essence, through enclosure and improvement, they stole the commons from the commoner (Rotherham, 2005, 2010, 2013b). As the old rhyme goes:

> 'The law doth punish man or woman
> That steals the goose from off the common,
> But lets the greater felon loose
> That steals the common from the goose'

It seems that today, in the twenty-first century, for all sorts of reasons, the geese are coming home to roost.

The upland bogs and mires

The loss of upland peatlands followed a similar track as the lowland fens though often driven by different pressures and opportunities (Rollinson, 1996; Rotherham, 2008b for example). Here the transformation was to sheep-moor and grouse-moor rather than to intensive arable farming. The social and economic drivers were different but the end result was the removal of common rights and usage and the shifting of local people from the land, desiccation followed and this was easier to achieve in upland sites since water simply drains downhill. However, the consequences for the downslope catchments have been very severe in terms of flooding and other events.

Wilder visions for future peatlands

It is clear from the preceding chapters, that whilst we are addressing some of the major problems that resulted from widespread destruction and degradation of peatlands, much remains to be done. Furthermore, in many countries we are now deeply entrenched in a period of enforced austerity in terms of public funding - of core local authority and agency services which deliver environmental works and nature conservation (Rotherham, 2015). This situation clearly limits the potential achievements of traditional conservation approaches. With this in view, and bearing in mind the broken nature of many ecosystems and their associated services, the search in on for a new way to reconstruct or mend a more effectively functioning ecology. Additionally, whatever solutions are proposed have to pay for themselves or be more economical than traditional approaches. Some innovations may involve changes to grant systems concerning land management and subsidies, though this may be harder in practice than in theory.

One umbrella concept which covers a diverse range of approaches is so-called 'rewilding' (Rotherham, 2014b). In essence this involves a variety of ways of reconstructing a more functional ecological system. It may involve strategic interventions such as re-wetting to trigger ecological recovery, or perhaps the reintroduction of lost keystone species to restore lost functions. Overall, the intention is to generate a less intensive landscape utilisation combined with a freer, wilder ecology. The question here is the extent to which such approaches are being applied to peatlands today or might be applicable in the future. Furthermore, what might the outcomes be for wilder and freer peatlands? It is important too, to recognise that the more extreme forms of free-willed nature and 'hard' rewilding may conflict with the need to recognise and to conserve the human cultural landscape and the archaeology or heritage features associated with that. Indeed, there may also be issues in terms of suggested intensification of 'bread-basket' farmland for food production and abandonment of other areas as 'wildlands', versus more widespread agricultural 'extensification'. The suggested approach is based on the assumption that the wildlands will deliver enhanced ecosystem services such as restored biodiversity. However, since much biodiversity and also cultural heritage depend on traditionally managed landscapes, and abandonment may trigger losses, then future initiatives should proceed with caution. This is particularly the case with peatlands and associated landscapes such as heaths and fens. Furthermore, in former peatlands now worked for intensive agriculture, the results of desiccation and regular ploughing mean that in many cases there are fewer than fifty harvests left in the soil. Release of carbon dioxide from such intensive farming is a significant contribution to global warming through greenhouse gas release. The cases for and against extensification and abandonment to wildland need to be more fully explored and assumptions carefully tested. Rewilding clearly offers major advantages and opportunities, but these are yet to be fully elucidated.

Finally, we live in a world dominated by issues of both natural and human-triggered climate change. It is likely that the massive removal of global peatlands over many centuries has been important perhaps even in triggering more rapid global warming. This topic and the precise impacts of peat removal can be debated, but the importance of both conserving what sites remain and restoring many of those that have been damaged is now widely accepted. Peatlands are globally significant reservoirs of sequestrated carbon, and furthermore, future sequestration will help offset more extreme changes. Additionally, peatlands in a good or at least an improving condition will help mitigate the undoubted changes that climate change will bring. The restoration of ecosystem functions and services will remain a key driver for future better husbandry of these under-appreciated landscapes.

A UK perspective

The rebirth of the English lowland peatlands

There are now ambitious plans to reinstate sites in the Southern Fens of England around Woodwalton (*The Great Fen Project*) and around Wicken, and in the Northern Fens centred on the Humberhead Peatlands and Potteric Carr (Rotherham, 2010, 2013b). These are exciting projects, though even at their best will nowhere near replace what we have lost.

However, the long process of recovery really began soon after the end of World War Two in the 1940s, when the reclaimed lands were under a new consideration for their future. Indeed, the plans for conservation and recovery emerged in parallel to the final throes of peatland removal and ultimately their conversion to modern, intensive, petrochemically subsidised agriculture. Alan Bloom's frustration on hearing that the land he had so recently won back from nature through the wartime efforts would be allowed to revert must have been extreme.

Conservationist Laurie Friday and her co-authors recounted a meeting between Bloom and a botanist walking on Adventurers' Fen (Friday, 1997). The botanist observed that it was "... such a pity that such a place .....should have to be destroyed", and how it would "....take years to grow up again into its natural state". This was a shock to Bloom who ventured that "... the thought the Adventurers' Fen might ever be allowed to revert had scarcely entered my head.... land, good land, now that it had been so hardly won back from dereliction, should never again be neglected or diverted from its true purpose".

So here we have the two diametrically opposed views, and each held with passion and a sense of righteousness. Indeed, it was these two views that essentially drove the bitter tensions between post-war agri-industry and the emerging conservation and environmental movements of the next fifty years. At Wicken Fen, a process of site and land acquisition has gone on since the 1950s with piecemeal proposals for restoration. However, by the 1990s, the National Trust was in a position to develop and deliver a management plan with a vision to encompass the wider site. Former arable land was brought back into pasture, meadow, or wetland condition with experiments to reintroduce lost plant species from seeds taken from the core relict sites. Is this a form of rewilding perhaps?

By the late 1990s, the Trust had restored to a wetter nature around fifty-two hectares of former arable land. This included their purchase of seven fields from Priory Farm which had been reclaimed for farming by Alan Bloom in the

1940s. In doing this they completed a cycle of acquisition which had begun in the late 1890s. Much of the reclaimed peatland had remained in agricultural production until 1994 and there are still problems with water amount and quality compared to that in a more natural landscape. Most of the site is hydrologically sealed behind banks and bunds of clay, but that does not necessarily ensure the required wetness.

There is a further issue with the loss from the fen of its long-standing traditional uses ('cultural severance'), and this is potentially very problematic. The National Trust managers do go to considerable effort to mimic former uses and create the required environmental conditions for specialist biodiversity. However, as Wentworth-Day pointed out back in the 1940s, the Fens were formerly at the heart of village life and as eco-cultural landscapes, (my words not his), the subsistence utilisation made them what they were. This was a socially and economically driven process and not one subsidised by grant aid and volunteer efforts.

The challenges remain very great. Indeed, these difficulties were alluded to in 1926 by one of the great champions of Wicken Fen, the eminent entomologist Frank Balfour-Browne (1926). He stated that

> Various efforts have been, and are being, made to preserve the fauna and flora in the small area of Wicken Fen, efforts which, I venture to think, will prolong the existence of the remnant, at most, for a few years. In spite of the steps which are being taken, the natural conditions are changing rapidly and with these, the [water beetle] fauna is also changing.

This gets to the core of the problem of changing conditions and the relatively small size of the remaining habitat areas – restoration is starting from a very low baseline. Furthermore, the problems he noted were getting much worse over the subsequent sixty years or so. It is only recently that there has begun a slow momentum of ecological recovery. Friday & Chatfield in the 1997 book on Wicken Fen finished with the following:

> We are sometimes asked what we are trying to achieve by all this activity at Wicken Fen. Perhaps the simplest answer is that we are trying to take the Fen backwards through time, to restore some of the biological diversity lost since the turn of the century. It is hard to read Alan Bloom's account of his struggle to drain and cultivate Adventurers' Fen or to watch the film he made of his efforts without admiration and heartfelt recognition of his excellent intentions. And yet we embark on undoing his work without regret: Britain is not at present in a state of food shortage and agricultural land all over the country is coming out of production. Now is an excellent time to attempt to restore lost wetland habitats and fenland species, and to recreate a landscape that has all but disappeared in 20th-century Britain.

Future fens

In terms of the environment, the landscape and its ecology and heritage cannot all be re-created; much is lost forever. Nevertheless, there is now a wider recognition of the imperative for peatland restoration in the Fens. However, despite the progress this is a long-term project and it is vulnerable to the vagaries and fluctuations in public opinion, the resolution or otherwise of agencies and politicians, and increasingly, the pressures of world markets for grains and fuel crops.

It has also become clear that whilst we can never reinstate the old fen, much of what we have today is not sustainable in the long-term. With a global environment of increasing food insecurity and more people, there will be a need to produce food, and indeed much more of it than previously. Much of this will of necessity be grown intensively and this is the reality of our world. However, there is a need for compromise for ethical environmental reasons and harshly pragmatic ones too.

Many of the former bread-basket fens are where food is grown intensively in countries like Britain, but there are complications. These landscapes need drainage but the crops often require irrigation as the areas are climatically dry. Furthermore, groundwater supplies are running short and many aquifers are at risk from contamination from, for example, seawater. Additionally, for farming there is the need to avoid the worst of the possible damage and disruption from both winter and summer floods. In many cases, these fens will not simply be wildlands but must be wetlands that once again work for a living and with local people. The new peatlands can be important catalysts for local economic renewal too, generating new crops and at the same time triggering tourism income and other related benefits. The old fen was an intimate mixing of nature and people that created a wonderful and rich environment and this is an aspiration for the new fenlands.

To bring about the necessary changes there must be a vision and that is something, which has been gradually emerging across this great English landscape. In the Southern Fen there are major projects centred on the Great Fen Project of Woodwalton Fen, the National Trust's visionary schemes for Wicken Fen, and the initiatives of the Royal Society for the Protection of Birds (RSPB) and others in Norfolk and Suffolk. For the Northern Fen, the bigger picture is yet to emerge although an increasingly coherent approach is gradually developing. In South Yorkshire for example, there are major projects at Thorne and Hatfield Moors and the Humberhead Peatlands as well as at Potteric Carr near Doncaster. But there is still no wider vision to restore the landscape on the scale necessary or to embed recovery within any plan to grow long-term economic gain. These new wet peatlands can be the pulse of regional ecology, economy, and community, and such a vision needs to grow from the most local level to a global transformation.

Examples of lowland recovery

Access and public engagement can be major barriers to progress – such as the unpromising gateway to Thorne Moors National Nature Reserve, England, which takes you over a derelict coal-mining site (Figure 14.2).

FIGURE 14.2 Access and public engagement can be major barriers to progress – this is the unpromising gateway to Thorne Moors National Nature Reserve, England, which takes you over a derelict coal-mining site.

The Great Fen Project

The Great Fen Project seeks to recover fenland over around 3,700 hectares between Peterborough and Huntingdon. The project takes a four-pronged approach to reconnecting the fragments of Woodwalton Fen and Holme Fen. The strategy is:

1. Implement restoration and habitat creation;
2. Improve access and enjoyment;
3. Foster socio-economic development;
4. Contribute to climate change adaptation and mitigation.

These are achieved by:

a A mosaic of wildlife habitats to support a variety of fen species,
b Provision of visitor facilities including a café, trails, natural play areas and events,
c New footpaths, cycleways, and bridleways linking to local communities,
d Areas to store floodwater, and
e Carefully zoning to allow access to some areas whilst protecting the most sensitive sites from too much disturbance.

Woodwalton Fen National Nature Reserve is the only one of the three sites in the Southern Fens where you can get a feel of a relatively undrained landscape.

The other sites are Wicken Fen and Chippenham Fen. Purchased by Charles Rothschild in 1910, and now owned by the Royal Society of Wildlife Trusts, it is one of the oldest nature reserves in Britain. In order to manage the site most effectively, the Middle-Level Commissioners and the site managers (the Nature Conservancy now Natural England) agreed to use the site as a flood-storage reservoir. Providing this important ecosystem service also helps keep the reserve wet by maintaining a higher water table than otherwise. A clay bank built around the Fen keeps water levels high, and in times of inundation floodwater can be stored on the reserve. For example, in 1998 the whole reserve of 208 hectares (514 acres) was under a metre of water. Yet the provision of this essential ecosystem service comes with environmental costs since the quality and quantity of water from such flood events damages the reserve's ecology. Since Woodwalton Fen is a Site of Special Scientific Interest, Special Area of Conservation, a Ramsar Site, and a National Nature Reserve designations placing a legal duty on the site's managers to ensure a favourable condition, a solution is needed.

Holme Fen might have been regarded as the fourth fragment of surviving ancient fen in Cambridgeshire. However, the site is cut into four sections by major agricultural drains which keep water levels well below what is needed and the famous 'fen post' records the effects. Fen and raised bog have given way to silver birch woodland, as the site is desiccating.

For these reserves survival and recovery depends on water tables and they are still locked into the issues of the wider farming landscape. Climate change adds potential problems in maintaining these sites with greater uncertainty for the future. There is an urgent need for imaginative, innovative solutions to help farmers make the best use of the land and to do so in ways that are sustainable in the long run. Current thinking is that climate change will radically alter rainfall patterns and thus be a further driver for the Great Fen Project and others like it. Indeed, with rapid climate change and regional urbanisation, Woodwalton Fen alone is unable to provide sufficient flood storage to alleviate the risks to infrastructure. This need for investment in flood storage capacity generates opportunities to develop additional storage and resolve long-term water-quality issues for the nature reserve.

Research has been undertaken to underpin the restoration with sound science with eco-hydrological modelling showing potential to create large areas of reedbed, wet grassland, fen, and wet woodland. Combined with engineering this approach can provide enhanced flood storage to ensure future flood-risk minimisation. An exciting possibility here would have been the re-creation of the lost Whittlesey Mere (drained around 1850), but sadly this is not possible; because of long-term topographic change due to the thorough soil wastage and subsidence, re-creation of England's greatest southern lake is not feasible. However, there will be some open water where the famous Mere once dominated the countryside.

Benefits from these peatland restoration projects include the provision of new countryside access areas where currently there is little existing provision, but increasing urban populations around the Cambridge growth-pole. Other ecosystem services provided with restoration and conservation management include the prevention of large amounts of carbon dioxide release from agricultural soils and new sequestration (capturing) carbon dioxide from the atmosphere into peat formation.

The Great Fen Project demonstrates how new approaches to fenland landscapes can deliver multiple benefits including socio-economic and environmental recovery. The intention is that land between the core reserves also returns to a mosaic of wetlands and other habitats alongside productive farming. This will help conserve remaining peat resources and create new floodwater storage areas. From this wider vision there will be a huge area of wildlife habitat available to be enjoyed by local people and tourism visitors. In order to achieve this transformation it has been necessary to acquire large areas of land. So since 2001, the project has purchased new sites and by 2008, supported by a grant of £8.9 million from the Heritage Lottery Fund, over 60% of the target area was now owned by the project partners. At that time, the grant was the biggest single heritage allocation ever given in England. The money enabled the purchase and restoration of land, along with the provision of education and community-based support for this previously undervalued but remarkable landscape.

For a long time, much nature conservation has focused on relatively smaller sites and protecting the best remaining places for wildlife, but the recovery of peatlands needs to be much bigger than this. Initiatives like the Great Fen Project are significant in that they change the scale of operation. The Great Fen Project is now one of the UK's cutting-edge landscape-scale restoration projects. It is successfully conserving internationally important wildlife and habitat and at the same time triggering major socio-economic benefits for the region and local communities.

The future conservation movement will need to think bigger and demonstrate the delivery of vital ecosystem services and local tourism economic benefit from wildlife-rich landscapes. Peatland restoration works such as the Great Fen Project are now showing how this can be achieved in practice.

Wicken Fen and the future

Returning to Wicken Fen, The National Trust and its partners have a strategy called 'The Wicken Fen Vision: Our Strategy to Create a large new Nature Reserve for Wildlife and People in Cambridgeshire'. This guides the future recovery of these once-great wetlands and has the necessary finance to turn the vision into reality. The location of Wicken Fen is important since it is only fifteen kilometres north-east of a rapidly expanding Cambridge. With emergence as a planned new town urban expansion, there is major pressure for new housing and other building development in the region. However, with the surrounding fenscape being dominated by some of the most intensively used productive farmland in the UK there are few opportunities for biodiversity or for outdoor recreation. Wicken Fen has survived over the last century as a virtual island in a sea of intensive agriculture. In the late twentieth century, nature reserve management here was focused on protecting rare species and intensive management of specific fen habitats. Nevertheless, some specialist species still died out and it has become clear that no matter how intensive the management, some of the species would be lost. The core reason was that the wetland reserve was too small and too isolated. From this observation there emerged the ambition to extend the site and enlarge it considerably.

With this in mind, in 1999, the National Trust launched its first 'Vision for Wicken Fen', to work with partners to extend the reserve sixteen-fold through acquisition of farmland to the south and east of Wicken. The site will provide landscape-scale outdoor space for wildlife and people, a 'green lung' for Cambridge, and will help climate-proof the region. Partnership and people are at the heart of this vision.

Much is being achieved in the Southern Fens but despite the ambition of these undertakings, the new sites are a tiny fraction of the size of the great fenland. The conservation sites are isolated islands in a sea of droughted, drained, and desiccated arable environment. Water always flows downhill to the lowest level and this remains a problem for a reserve like Wicken where site re-creation is restricted by surrounding land use. The restored area gives a sense of the old fen, but much of it is dry and botanically species-poor, and without wider landscape-level management of water it will stay so. The remains of the old fen are more secure in their bunds and embankments, in part buffered by the new lands, but still an isolated, embattled fragment. The future challenge will be to re-wet the wider countryside so the expansive nature of Wicken becomes a reality.

Restoration and re-creation in the English Northern Fenland

The Humberhead Peatlands National Nature Reserve is at the heart of what was the great Northern Fenland. Much of this landscape has been researched by scholars such as Paul Buckland (e.g. Buckland, 1979; Buckland & Dinnin, 1997), and there are substantial volumes produced and edited by Robert van de Noort for example (e.g. Van De Noort & Ellis, 1997). Yet despite this, in terms of site management and restoration, the archaeology and heritage interests are generally overlooked and ignored. Bog 'oaks' removed from the peat mass are simply piled up on adjacent farmland and, undocumented, they are left to decay. For many years, one of the battles over Thorne Moors was to gain recognition for the need to protect and conserve the 'palaeo' environmental record which somehow fell between the 'natural' environment and the geological resource. Consequently, it was ignored and unprotected; there is a lesson here for peat restoration projects around the world.

The reserve today includes Thorne Moors, Hatfield Moor, Goole Moor, and Crowle Moor totalling 2,887 hectares and forms the most extensive restored area of former raised bog in lowland Britain. The site is a Special Area of Conservation (SAC) for its peatland habitat and is an internationally important breeding site for the nightjar (*Caprimulgus europaeus*). Natural England intends to establish the Moors as 'centres of peatland excellence' with visitor facilities to match, but visit today and you realise the task ahead. The challenge is firstly to find the site and then to access it. Once you have done that then the area is stunning with diverse habitats supporting 5,000 or more species of plants and animals including over 4,000 insect species recorded. The herptiles are good too with a nationally significant population of adders. Over 200 bird species have been recorded with around seventy-five species breeding and winter-visiting whooper swans (*Cygnus cygnus*), pink-footed geese (*Anser brachyrhynchus*), rough-legged buzzards (*Buteo lagopus*), hen harriers, and short-eared owls. Summer visitors include woodlark (*Lullula arborea*), oystercatcher (*Haematopus ostralegus*), lapwing (*Vanellus vanellus*), ringed plover (*Charadrius hiaticula*), great crested grebe (*Podiceps cristatus*), and at its most northerly English site, the nightingale (*Luscinia megarhynchos*) though this may have been lost. Interestingly, this declining species was formerly associated here with extensive beds of invasive *Rhododendron ponticum* which have since been cleared. In recent years, evoking the image of the old fen, common cranes (*Grus grus*) have bred.

The Humberhead Levels Nature Improvement Area and National Nature Reserve

The Humberhead Levels Nature Improvement Area (NIA) was declared in 2012, as one of twelve such specially funded projects in England. The project was kick-started with a grant of £587,295 and additional support to the tune of £1,034,760. The area is part of the vast flatlands of the former great Northern Fen which straddled Yorkshire, Lincolnshire, and Nottinghamshire.

It is suggested that the area offers the finest opportunity in England to develop a major multi-functional wetland landscape in a hitherto largely unrecognised biodiversity hotspot. The NIA extends across nearly 50,000 hectares of the Humberhead Levels National Character Area (the UK government's way of identifying and classifying distinctive landscapes). Interestingly, in terms of peatland restoration this is a different wetland model from many others applied elsewhere. In this case the habitats are intimately interspersed within a matrix of highly productive arable land. Most of the area is close to or below sea level and highly vulnerable to climate change impacts. The main aim is to create an internationally renowned, unique wetland landscape, supporting thriving communities, economy, ecosystem services, and wildlife. Sadly, even in these new initiatives there is still no obvious recognition of the globally significant palaeo-heritage of unique timeline profiles of these peat bogs and the fossil forest of ancient trees. I suspect that in say the USA the area would have long-since been declared a 'National Monument'.

This project is not about turning back the clock to when the Levels were an impassable wetland and an ancient, intractable swamp and multi-domed raised mire. Rather it will enhance existing internationally important wetlands (the Humber and the Humberhead Peatlands), the other SSSIs, and Local Wildlife Sites. It is hoped that these sites will be reconnected by working with local farmers to create ribbons of habitat on unproductive drain-sides, headlands, and wet field-corners, and these will mesh with the important rivers and dykes that cross to the wider area. At least, this is the intention with wildlife free to move through adjacent farmland; land economic values will be maintained and climate change resilience increased. Work will create or restore nearly 1,500 hectares of key wetland habitat which will evolve over time into peatland. In essence one output will be progress towards the reinstatement of England's largest lowland mire system; though to be fully functional may take around 500 years. As in the Southern Fens, successful conversion to wetlands here will increase carbon sequestration into newly forming peat and wetland soils, along with other ecosystem service provision. Of course there is a huge challenge to cross the gulf between aspirations and actuality.

Importantly, the NIA which was funded for less than five years will hopefully have triggered the development of community capacity to get involved with these wildlife sites, something sadly lacking previously. The approach had three distinct facets:

1. First, there was improved interpretation and face-to-face contact with visitors on the five sites which already had visitor infrastructure. It was hoped this would grow to 6,000 additional visits to local wildlife sites over the early period.
2. New environmental education programmes operating from three sites were to improve awareness and community engagement.
3. There was targeted volunteer development and training to deliver an extra 3,910 hours of volunteer input to the projects, and hopefully the enthusiasm generated will carry the work forward beyond the grant-period.

Improved integration of land use is intended to enhance climate change resilience. Additionally, closer partnership working is aiming to align farming with sustainable flood defence, water supply, and nature conservation. There is also the aspiration that the local 'green economy' will grow through work on initiatives such as energy biomass projects. Furthermore, it was suggested that such investment might provide new jobs and sustainable development opportunities in areas of 'multiple deprivations' like Thorne Township, though it is unclear how this would come about in practice. The projects provided some employment through conservation management contractors and opportunities like green tourism employees such as café staff or on-site nature guides.

The expected success of the project lay in the hope that connecting local communities to their wildlife sites would increase local pride and stewardship and so would grow the independence of the initiatives from central government funding. This would help foster social well-being and provide significant local health benefits, but without core financial support this may be a somewhat naive belief. Whilst these aims are laudable and achievement would help the long-term recovery of some of the degraded peatlands, there are still major problems. Like many damaged peatlands the core region lacks any main tourism attraction with the necessary facilities to draw visitors in and be an economic growth-pole. Indeed, the initial target of 6,000 new visitors was somewhat trivial in terms of any significant economic impacts due to tourism and was not effectively audited. Despite the stated aspirations, the area remains a tourism desert and is without any joined-up tourism visitor strategy. This is an important reality-check for peatland restoration projects generally if they put forward economic and tourism impacts as justifications for funding leverage.

The few visitors who do come to Thorne and the surrounding areas spend little or no money to spare in for the local economy. Essentially there is nowhere to spend it and nothing to buy. In this way, the area is trapped in a vicious cycle because the unemployed and economically deprived local community see little tangible benefit attached to the conservation site. Worse still, the recent influx of 'outsider' visitors to the area may not be appreciated by local disaffected residents. These are difficult issues and need to be addressed if long-term recovery of some of the more economically remote peatlands is to be successful.

The core sites within the Humberhead Levels include Thorne Moors itself, Hatfield Chase, and Crowle Moors (managed by the Lincolnshire Wildlife Trust). Thorne and Crowle, having always retained significant water, are subject to careful water management and restoration schemes. Thorne Moors itself has had a carefully implemented regime of very gradually raising the water levels. and this is necessarily a long, slow process. A higher water level could be easily achieved but would fast-track the site to open-water and reed-fen, and not sphagnum bog. The latter is the ultimate aim of restoration but the key vegetation is easily eroded by wind and waves across open water; this is a problem that has been experienced at nearby Hatfield Chase, for example. There are transferable experiences here for other lowland restoration sites in that it is quite easy to create a wetland but much less easy to generate a long-term peatland.

Potteric Carr nature reserve

Potteric Carr was the greatest and most intractable of the Yorkshire Carrs and a peatland site described in the late 1700s as unfathomably deep (see Rotherham, 2010). Historically, until the site was drained in the 1700s and 1800s it was mostly bog, carr, and fen, but became agricultural after drainage. The area was then affected by Victorian coal-mining and was crossed by the mainline railway. During the mid-1900s, mining subsidence occurred under a part of the Carr and within about twenty years, the land began to revert to something resembling its former fen and carr state. Associated plants, birds, and insects began to reappear or return, some probably surviving in the drains which criss-crossed the area. In 1968, a small area of the Carr was designated a nature reserve by the Yorkshire Naturalists' Trust and this was expanded in the mid-1970s. However, in recent decades the site has been increased massively to attract breeding bitterns (*Botaurus stellaris*) to newly created reedbeds and with improved visitor facilities. The area has had a chequered history of industrial impacts and then creeping urbanisation as Doncaster has expanded.

Following 1970s proposals for further railway lines across the site, the Potteric Carr Internal Drainage Board was examining the possible use of the Reserve for flood relief for the south Doncaster urban expansion. This was perhaps the first realisation of potential ecosystem services which could be provided by the expanded reserve. The proposed flood-relief scheme was intended to relieve pressure on downstream river systems and was modified to address both conservation and flood-relief issues; construction from 1979 to 1981 brought the system into use with land drains regraded and deepened, and two new pumping stations. The site acted as a store for floodwaters which were released later when any flood surge subsided. The scheme was to recover the wetness across the wider site following the progressive drainage from the 1700s. The changes resulted in a lowering of the water table over the whole area and a groundwater-dominated regime changed to a perched water table. However, during winter with normal rainfall all the water bodies filled naturally and the clay substrate retained water on-site. Summer water levels dropped through evaporation and transpiration but water from the Mother Drain, the only source of water throughout the year, could be extracted by permission. More recently, development of housing, light industry, and leisure on land to the north and west of the Reserve has meant an increase in flood flows. This is because of farmed pasture being converted to hard surfaces for urban development and increased volume and speed of run-off. In response to this situation, there has been a major expansion of the nature reserve to provide both flood-control and 'water-polishing' through the created reedbeds. These works were funded through 'planning gain' from the development. There was also a major investment of around £2,000,000 from the Heritage Lottery Fund and the EU which was for the creation from former farmland of seventy-five hectares (185 acres) of new reedbeds for bitterns.

The result of these actions as described briefly here has made Potteric Carr into a 320-hectare (791-acre) area nature reserve of low-lying wetland only two miles from the bustling centre of Doncaster. Breeding and wintering bitterns and little bitterns have been highlights of a major recovery of wetland biodiversity. More than seventy different bird species regularly breed on the reserve and there is a rich diversity of insects particularly butterflies, dragonflies, and damselflies. Mammals include water voles (*Arvicola amphibius*), bank vole (*Myodes glareolus*), grey squirrel (*Sciurus carolinensis*), brown hare (*Lepus europaeus*), badger (*Meles meles*), roe deer (*Capreolus capreolus*), and various bats.

In many ways, Potteric has come through difficult times to emerge as a success story for wetland conservation. However, it has lost its old traditional uses and these will never return. Nevertheless, for a site that suffered so many setbacks to have re-emerged as it has, this is a very positive outcome. The site also indicates potential new uses and new ways to connect such wetlands to local people and the economy. There are around 25,000 or more visitors annually and this boosts the local tourism economy. Furthermore, Potteric has been shown to be good for business as a high-quality visible green gateway to the emerging centre of Doncaster. By establishing the carefully designed series of interconnected reedbeds, streams, and lakes, a major function of the site is to provide a suitable breeding habitat for rare birds such as marsh harrier, bearded tit (*Panurus biarmicus*), and bittern. However, the reserve also cleans wastewater, 'polishing' effluent from 25,000 Doncaster homes and providing more than 200 'Olympic swimming pools' worth of floodwater storage. This is a vital flood protection for farmland, businesses, homes, and transport infrastructure.

Whilst Potteric Carr is clearly a remarkable success story and certainly delivers some of the wider benefits that the Thorne Moors sites seek but presently miss, there are issues. The main one is that whilst a fen peat will be slowly developing, the site is presently a wetland but not a peatland. A mix of carr woodland, reed fen, and open water may ultimately generate a mire ecology but it is likely to be a very long wait. What Potteric demonstrates well, and is transferable to most if not all peatland restoration projects, is the importance of managing water. However, the site is also caught by a dilemma that in delivering the valuable ecosystem services such as wastewater polishing, it is trapped in a highly eutrophic ecology which is not (and never will be) conducive to mire recovery.

Other significant fenland restoration in the UK includes the Somerset Levels and Moors project, the Anglesey and Llŷn Fens LIFE project in Wales, the Lakenheath Fen project in East Anglia, the Lancashire Mosslands project, work by the Cheshire Wildlife Trust on the meres and mosses through their living landscapes project, and a little further south, the Marches Boglife project. This is restoring sites across Shropshire and the surrounding areas. In 2016, the Marches Mosses Boglife project gained £5 million Heritage Lottery Fund and Life funding to continue the recovery of Fenn's, Whixall, Bettisfield, and Wem Mosses, which combined are the third-largest lowland-raised peat bog in the UK.

Restoring bogs and mires

Just as in the lowlands of Britain there are major opportunities to begin the process of long-term recovery and reconstruction, so too are there initiatives and potential projects in the upland and northern areas. However, this is the start of a process rather than the quick fix of a conservation 'product'. With degraded peatlands anywhere in the world, restoration is a long-term goal, and in many cases the created sites will for a very long time be wetlands but not peat-based. For projects and programmes of peat restoration nationally and internationally, there is much to be learnt from the works that are already underway. Many examples have now been published (e.g. IUCN reports and particularly Brooks & Stoneman, 1997; Cris et al., 2011, 2014; and on individual project websites etc).

In the upland and northern regions of Britain, restoration projects are focused more on bogs rather than fens. The following examples are from Cumbria, the North Pennines, the South Pennines, and Exmoor.

Cumbria Boglife project

With funding from LIFE+ EU Cumbria Boglife has been working with Natural England to restore lowland raised bog at Bolton Fell Moss, South Solway Mosses, and Roudsea Wood and Mosses. There has been a lot of wider community engagement and awareness-raising activity as well. The project has aimed to reduce past damage from past activities and ecological successional processes that threaten the peatland's survival. The key issues addressed include woodland encroachment, scrub invasion, invasive species, past peat extraction, and land-drainage for both agriculture and forestry.

The specific activities funded over a five-year period were:

1. Restoration of 507 hectares of damaged lowland raised bog within three sites in Cumbria: (a) Bolton Fell Moss Site of Community Importance, (b) South Solway Mosses Special Area of Conservation (SAC), and (c) Roudsea Wood and Mosses SAC;
2. Use of the sites to demonstrate a range of restoration techniques to managers of similar habitats and wider audiences;
3. Monitoring of the recovery processes and sharing of best practice guidance about the restoration techniques;
4. Raising awareness about the importance and value of the sites and lowland raised bog habitats in general. The latter was done through events, education programmes, newsletters, a dedicated website, and social media, together with mid-term and end-of-project conferences.

Yorkshire Peat Partnership

Yorkshire has around 20% (70,000 hectares) of England's remaining blanket mire and a further 100,000 hectares of degraded upland peats that no longer support characteristic bog vegetation. This, of course, does not include areas historically stripped of blanket peat and converted to farming long ago. However, this does clearly suggest the scale of the challenge to turn the remaining bogs into active peat-forming condition, a task taken on by the Yorkshire Peat Partnership. The partnership established in 2009 is managed by Yorkshire Wildlife Trust and its original aim was to restore to good condition 50% (around 35,000 hectares) of Yorkshires existing blanket mire by 2017. The partnership includes major stakeholder organisations such as the National Parks of the North York Moors and Yorkshire Dales, the Environment Agency, Natural England, Yorkshire Water, and the National Trust.

Conditions assessments suggested that the reasons for vast areas of the county's peatlands being degraded were:

1. Changing climate;
2. Historic atmospheric pollution;

3. Over-grazing;
4. Wildfires and poorly-managed burning;
5. Recreational erosion such as along footpaths.

The partnership set about addressing these issues through a mix of remote sensing assessment, field surveys, and mobile GPS-enabled computers which allowed strategic planning of actions. Following detailed site survey and assessment full restoration plans are drawn up. The level of specific detail makes it relatively easy to draw up implementation contracts and to estimate labour and materials etc.

The surveys indicated 5,850 kilometres of grips needing to be blocked and with these, around 1,768 kilometres of gullies and hags associated with 340 hectares of bare, badly eroding peat. Within two years of the project inception there were 16,542 hectares surveyed, 3,283 hectares under restoration works, 334 kilometres of grips blocked, forty kilometres of eroding gullies revegetated, seventeen hectares of bare peat revegetated, 33,000 peat dams installed into grips, 300 sediment traps into grips and gullies, and much more. This is a huge achievement and by the end of the first phase of funding there was a considerable amount of work completed.

The Yorkshire Peat Partnership worked with a range of key stakeholders and this was probably a key message from the project. Being able to pool and share vital resources, skills, and enthusiasm helped coordinate activities, develop new and innovative technologies and solutions, and essentially overcome major barriers and problems. The outcome was to achieve peatland restoration on a scale that would have been impossible for any single organisation.

Moors for the Future and the MoorLIFE restoration project

Further south but also partly in Yorkshire in the South Pennines and Peak District, the 'Moors for the Future Partnership' also addressed blanket mire recovery with a long-term recovery programme. This has been based around the massive upland landscape of Bleaklow in the Dark Peak. With a huge amount of medieval peat removal from the Dark Peak moors and surrounding areas, and then catastrophic air pollution from industrial cities over perhaps 200 years, wildfires and intensive sheep-grazing, the landscape was reduced to a vast area of bare and eroding black peat. Furthermore, the ecosystem services of water management (quality and quantity), biodiversity, and carbon sequestration were reduced to almost zero. Repairing this long-term damage was at the heart of the project funded by EU LIFE monies with the intention to restore over 800 hectares of degraded bog by 2015. The project has continued to address the issue since that time and has applied both novel technical solutions and citizen science survey support to achieve its targets.

One of the key aspects of this project was in attracting private sector investment from Yorkshire Water plc to deliver major ecosystem service recovery. The main deliverable here was clean water and not brown, peat-stained water that customers don't want and which cost the company millions of pounds per year to remove from water supplies. By restoring the peat surface to vegetation and by blocking groups and gullies, the project has made excellent progress in the long-term solution to such problems.

One approach was for lime and heather brash to be applied to the bare peat surfaces and a short-lives grass-seed sown to provide stabilisation and cover. Gulley blocking and geotextiles were also used. However, one of the major innovations, though not totally without controversy, was to introduce sphagnum mosses by either 'beadamoss' pellets (with tiny sphagnum plants inside a gel ball), or by plant plugs. The controversy amongst specialist ecologists and naturalists was that there was no effective baseline survey of prior conditions and status of sphagnum mosses on the receiver sites or in the surrounding landscape. Whilst many of the target areas were indeed devoid of moss, there are colonies around the sites and tiny sphagnum spores can easily recolonise if conditions become suitable. However, such a longer-term approach does perhaps take more time. The other issue has been that there appears to be limited recording of the sphagnum species and proportions introduced into each site, their provenance, and subsequent survival. There is always a risk of the transmission of diseases and pests if materials are outsourced and also this feels like a missed opportunity to gain more insight into the survival and establishment processes.

Despite these reservations the work has been successful, though there are still issues of effective establishment of sphagnum on the more extreme sites. Furthermore, any real return to a fully functional mire system will naturally take a very long time. In such extreme conditions recovery is painfully slow but decreasing erosion and holding back water can be fairly quick. Re-establishing the mire hydrological balance is far more difficult.

The citizen science work has generally been exemplary in terms of engaging local communities and training students and others in important field skills. However, again there are some niggles and this applies especially to the online and downloadable app to help record and identify sphagnum mosses. Whilst this helps engage local people and involve them in surveys, the data generated are really limited to the presence or absence of 'sphagnum' generically. Although the sphagnum mosses are a keystone contributor to the ecosystem here, their identification to species is notoriously difficult. The app is seriously misleading and does cause beginners to think they have found rare species which in practice have not been within hundreds of miles of the region for many decades. This is rather unhelpful and you simply cannot get meaningful data with single online images.

Overall, the project has been hugely successful with major improvements in water quality, and in carbon capture for landscape, flood mitigation, and biodiversity. Being in the Peak National Park and on the Pennine Way, the region also delivers major local economic benefits through enhanced tourism.

Exmoor Mires project

The Exmoor Mires Project has been a very carefully planned and executed restoration programme for over 2,000 hectares of upland moor and bog. Led by South

West Water plc's Upstream Thinking Catchment Management Programme the project has engaged key stakeholders such as the Exmoor National Park, The Environment Agency, Natural England, and private landowners. The network has operated through the Exmoor Mires Partnership which was formed back in 1999. Approaches developed elsewhere in the UK to deal with issues like grip blocking have been adapted to the thin peats and relatively steep slopes of Exmoor.

One innovation by the Exmoor project has been to address at least in part, the issues of landscape heritage and archaeology on these culturally rich landscapes (e.g. Bray, 2015). With the known cultural heritage and significance of Exmoor the project commissioned detailed landscape heritage and archaeological assessments. Not only did the project assess archaeological interest prior to physical interventions on-site, but they even went some way towards evaluating peat-cutting archaeology. The known heritage interest was then both documented and taken into consideration in restoration plans. By 2010, the project had blocked over fifty kilometres of ditches which helped re-wet more than 350 hectares of moorland.

There are numerous other major UK bog restoration projects, far too many to mention them all. However, these include work at Malham Tarn in North Yorkshire, Langlands Moss in South Lanarkshire, Scotland, the Sustainable Catchment Management Plan (SCaMP) in north-west England, the North Pennines AONB Partnership Peatland Programme in northern England, May Moss in the North York Moors National Park, the Cuilcagh Mountains in Fermanagh, Ireland, Lake Vyrnwy in Wales, and the Flow Country in northern Scotland; there are several others, and together they make a massive contribution to redressing some of the damage done over many decades.

Some lessons from the uplands

The wider range of current projects on upland peat restoration has promoted many different approaches. Furthermore, the various projects provide opportunities to test the methods and to gauge success. To this end, the publications such as by initiatives like the IUCN Peatland Programme mean that much good information is available to be used by other practitioners and projects.

There are significant differences between upland and lowland peat restoration projects in terms of environmental conditions, history of exploitation and destruction, and topographic context. Essentially, these are higher altitude sites with cooler, wetter weather and longer winters, low levels of nutrients in both soils and water, and are socio-economically peripheral areas. Whilst they have often had human impacts over thousands of years (as early settlement often began in upland zones), the landscape is not amenable to intensive utilisation. Indeed, prehistoric ploughing and cultivation probably triggered massive erosion, downwash of sediment, and acidification of skeletal mineral soils. Whereas in lowland fens the incentive for land 'improvement' and drainage was highly productive arable land, this was never the case for the uplands. The end usage has often been sheep moor, grouse moor, water catchment and sometimes forestry.

Upland areas were relatively easy to drain in terms of simply cutting downhill and allowing water to flow away, whereas lowland sites often needed pumps. Furthermore, unlike the lowlands, in order to drain off a hill, you don't require political or economic control over the surrounding ground. So whilst upland sites were easier to drain, they can also be simpler to re-wet although sloping ground denuded of peat may take a long time to re-vegetate to bog vegetation.

A final lesson for the uplands is that the major driver to recover peat bog status is the financial cost of water supply in terms of both quality and quality of water. Not only do the peatlands here deliver hugely important ecosystem services, but there are also direct economic implications when these services fail. There are also climate-related services such as carbon–fixation and downstream flood-risk too, but the economic trail for these is less direct.

A wider context

Europe

Across the continent there are around 505,000 square kilometres of peatland remaining or about 13% of the known global resource. Furthermore, because of the long history of intensive human settlement and exploitation, many of these areas have been badly compromised and others (unaccounted for) have been destroyed. Extensive areas occur in Finland, Sweden, European Russia, Ireland, and formerly Holland, Germany, and France. It is recognised that "Peatlands are a critical element of Europe's natural capital and investing in them as part of Europe's green infrastructure will not only significantly contribute to EU biodiversity goals, but will also provide multiple ecosystem benefits to society in the face of climate change". This was stated by Micheal O'Briain as deputy head of the Nature Unit in the European Commission (in Cris et al., 2014). Yet we know that agriculture, peat-mining, and forestry have turned many parts of Europe into a major hotspot for global carbon dioxide emissions.

In this context there are numerous projects to begin the process of restoration, and indeed to halt ongoing further losses. Just a few of these include the following:

Irish raised bog restoration by Bord na Móna

Bord na Móna has been the Irish government agency which destroyed huge areas of peat bog for energy generation in the post-World War Two period. However, several formerly raised bogs were identified in the 1980s as still having potential for conservation-restoration and these have formed the core of the agency's programme of reinstatement. The total restoration was around 400 hectares of formerly raised bog by 2014 with other sites coming on stream. The approach used was to undertake detailed topographic surveys and then undergo a programme of drain blocking.

Sustainable utilisation of Belarus wet peatlands

The Wetland Energy Project has helped divert a factory manufacturing peat-fuel briquettes, into biomass energy generation from sustainably managed wetlands. With 1.5 million hectares of peatland in Belarus, around 50% has been drained for farming. By 2014, 50,000 hectares of damaged peatlands had been re-wetted and there was the possibility of re-wetting a further 0.5 million hectares. Expected benefits of the project included socio-economic growth, that is, jobs, in remote rural areas as peatland exploitation had declined with peat exhaustion, reduction in peatland fires over drained and desiccated bogs, and recovery of biodiversity interest.

Re-wetting Brandenburg's Alkaline Fens

This one-time common habitat in north-eastern Germany has been decimated by changes over the last two centuries. A recent project set out to restore fourteen of the previously damaged sites. Like the English fenlands described earlier, the countryside here was drained and turned into a profitable but unsustainable intensive arable farming. On abandonment, the fenland successions were badly affected by changed nutrient levels and there was degradation and mineralisation of the peat soils. The project involved not only the blocking of the extensive drainage networks but also the reinstatement of some degree of traditional farming on formerly managed but now abandoned fens. This latter approach involved close liaison with local farmers.

Peatland biomass for fuel at Biebrza, Poland

Poland has some remaining extensive peat bogs but around 84% of the known peatland resources of about 12,500 square kilometres are degraded, mostly used as hay meadows and pastures. The initiatives here include integrated management to connect peat restoration with economically viable farming and water management. An iconic bird species, the aquatic warbler (*Acrocephalus paludicola*) has been used as a flagship species for wider conservation efforts. Poland holds 25% of this formerly widespread but now globally threatened passerine. The bulk of this population is in the vast peatlands of Eastern Poland and it is here that efforts are being concentrated.

From restored wetlands, the products included energy biomass for biogas production. In order to maintain the necessary habitat for the warblers, the projects depended on appropriate agricultural subsidies for local farmers.

Sphagnum farming or paludiculture on formerly degraded German peat bogs

Important in the long-term restoration, recovery, and conservation of peatlands is the development of appropriate economic drivers for these aspirations. Unfortunately, simply establishing major ecosystem service values is not enough in a society in which politicians blanch at the idea of long-term investment of public monies for long-term public benefits (see, for example, Rotherham, 2015). It is therefore important to have real-time economic values that support conservation and this can be a challenge.

However, considerable progress has been made in recent years to both restore historically damaged bogs and also to find economically viable uses for the countryside. One very positive initiative has been to use sites to grow sphagnum moss to produce peat-alternatives for horticulture and also moss for hanging baskets and other plant culture. This use is called 'paludiculture' and has been pioneered in Germany. Methods developed so far include land-based cultivation and water-based with artificial floating mats. Work is now also being done in the UK.

Biomass from re-wetted German peatlands

With drained peatlands turned to agriculture and incipient climate change it is necessary to find viable alternative uses for these sites if re-wetted. One alternative usage for a re-wetted German peatland has been the production of biomass to replace oil in electricity generators. The Pomeranian peatlands cover about 12% of the land area in Mecklenburg-Western Pomerania and with this landscape drained for agriculture has been releasing 27% of the total carbon emissions from the entire province. Some of the farmer stakeholders in the area have turned to re-wetting the landscape and exploiting it for biomass and this is commercially and environmentally viable.

Peatland restoration in Sweden

The Swedes have been restoring around thirty-five degraded mire and wetland sites across the country, with a total area of around 3,000 hectares of peat bog and fen. The intention has been firstly to restore hydrological functioning, and secondly to bring back some traditional farming such as haymaking, meadow mowing, and reed management; they hope to reduce carbon flux and improve water quality as well. One problem experienced about the geography of Sweden and transferrable to some other countries, was the long distance between the restored peatland sites and road infrastructure.

Africa

As discussed earlier, the peatland resources of Africa are still only poorly known. Estimates range from around 5,000 square kilometres to 13,000 or more, which means this could equate to nearly 3.5% of the global resource. Some of the sites are incredibly old and have been unaffected by things like glacial periods. The South African Mfabeni Mire, for example, has a depth of up to twelve metres of peat and is estimated to be around 45,000 years old. However, in regions such as East Africa, peat-mining has resulted in regional carbon emission hotspots.

South African wetlands

Wetlands in South Africa are generally limited by high evapotranspiration, often dependent on groundwater, and are limited in extent. Sites like mires and other peatlands, where they occur, have been extensively exploited for agriculture with dams, drains, cultivations, use as pasture lands, and severe overgrazing. The result is that the fens are eroding and the drained peat-swamp forest is converted to banana plantation, eucalyptus timber forestry, and irrigation projects. Peatlands then dry out and are often burnt. In recognition of the losses in ecosystem services, the South African National Biodiversity Institute established the Working for Wetlands project. The vision includes finding much-needed employment for the poor in these locations. Funded by around $63 million the initiative created up to 15,000 jobs and set about undertaking work to prevent erosion, rehabilitate the wetlands, and use engineering solutions to restore degraded sites. There is an emphasis on job creation, capacity building, partnership working, communication and education, and associated research and planning. Around 40% of the sites targeted are peatlands.

The Rugezi Mire, Rwanda

Rugezi Marsh is located in northern Rwanda, east of Lake Bulera. This is high-altitude mire and is naturally formed by a floating mat on a layer of peat that is between seven and twelve metres deep. However, the site has been compromised by hydro-electric projects in the lake and increasing demand for water supply to a growing local population. Water levels in the lake have been reduced critically so that they threaten the hydro-electric scheme and over 50% of the marsh was degraded by farming activities. However, drainage and cultivation have been stopped in the worst affected areas of the marsh and there has been a remarkable recovery of both water quality and water quantity. Interestingly a non-native tree (*Alnus cuminata*) has been introduced to form a conservation buffer around the core marsh zone and is used for fuelwood, construction, and day-to-day items such as broom-handles and poles. Part of the issue in recovering the site is the importance of the landscape to local subsistence and there are major socio-economic challenges ahead in order to recover the ecosystem service functions.

Asia

A vast continent, Asian 'Russia' alone holds around 1.2 million square kilometres of peat which is about 75% of all the peat resource of Asia. Indonesia has around 17% and the remainder is in China, Malaysia, and Mongolia. Asian peatland habitats range from the Arctic and sub-Arctic tundra, to tropical swamp-forest, and temperate sphagnum bog. Some sphagnum peatlands such as in Georgia are relatively unaffected by human impact, whereas in neighbouring Turkey, for example, there are just a few sites and the fen peats, for example, have been heavily exploited by cutting.

Many of the tropical Asian peat sites have been hugely degraded by drainage and burning and been converted to agriculture or plantation forestry (as described earlier). In China the plateau peatlands are badly affected by both cutting and overgrazing by livestock. Temperate and tundra peatlands in Russia are in part buffered by their extent and by extreme climatic conditions but they have been affected directly by peat-mining and indirectly through activities such as oil exploitation.

The Ganges delta in Bangladesh

Water is at the heart of Bangladesh and the country has about a quarter of a million hectares of peatlands. Furthermore, Bangladesh is very poor and has a rapidly growing human population, a recipe for major environmental problems. The peatlands are mostly situated between the Ganges river floodplain and the Ganges tidal floodplain. The resource has been degraded by the extensive use of peat-fuel in the traditional cooking stoves of the poor people, by harvesting of snails from the organic sediments to feed commercial shrimp farming, drainage and reclamation of peats as arable farmland, and pollution from fertilizers and pesticides washed off by the nearby agricultural field systems.

The work up to around 2014 consisted mainly of working with local farmers to raise awareness of the importance of the peatlands and surveying the area to assess the current status. This will hopefully lead to a strategy to halt the decline and begin a process of restoration.

Mongolian peatland

Inner Mongolia is climatically a dry region but the extensive Wuliangsuhai Lake is a major water-body extending over around three hundred square kilometres. With surrounding agriculture the water sources for the lake have been badly affected by the run-off of fertilizer (causing eutrophication), pesticides, and other contaminants. Productivity in the shallow (one-metre deep) lake historically has meant the accumulation of peaty organic sediment. This project is really concerned with a wetland more than a peatland as such, and the growth and harvesting of common reed (*Phragmites australis*) helps lower the elevated nutrient levels. The reed is used as fodder, for making mats and other items, and for pulp and paper production.

China's Ruoergai Plateau

On the border of Sichuan and Gansu Provinces, this region has nearly 5,000 hectares of peatland associated with the upper catchment of the great Yellow River. However, overgrazing by livestock has led to the lowering of the water table and there are drains and resulting gulley-erosion across the area. The problems are compounded by climate change with rising temperatures affecting the alpine flora.

The peatlands here provide various major ecosystem services including water supply, carbon storage, biodiversity, agricultural grazing, and potential tourism. Conservation intervention has therefore sought to safeguard and restore the condition of the bogs. The main actions have been to block key drains and hold back water, to fence areas to reduce erosion from yak grazing and trampling, and attempts to re-vegetate damaged areas. From the efforts so far, over 1,500 hectares of peatland has been successfully restored. Demonstration sites have been used to show the benefits of restoration, and an improved understanding has helped refine the conservation approaches employed. However, the long-term challenge will be to address the region's socio-economic issues and find alternative income-streams to support the farmers whose income presently depends on grazing drained and degraded peatlands.

Indonesian peat-forest

As discussed earlier, peat-forest degradation in Indonesia is a major problem. The country holds one of the greatest extents of peat-swamp forest in the world but is desperately poor and with a growing human population. Extensive peat forest occurs on the islands of New Guinea, Sumatra, and Borneo, but in recent decades the latter two have lost around 70% of the resource. Furthermore, ongoing destruction is displacing local indigenous peoples and combinations of drainage and deliberate wildfires have made Indonesia a global leader in carbon release linked to land use. Drainage alone is responsible for around 500 million tonnes of carbon dioxide released per year. Oil palm and pulp plantations already extend over about a fifth of the west Indonesia peatland, and it is indicated that the peat-swamp forest will have gone from that area by 2030.

Restoration programmes depend on finding alternative economic crops that can be cultivated in peat-swamp habitats. This is combined with attempts to re-wet the landscape and to prevent further drainage. As so often the case, the long-term solution will lie in educating local communities about the importance of the forest and at the same time empowering them to take appropriate action. A complication in Indonesia has been the intervention of high-level corruption with economic interests in government and private businesses that have driven much of the destruction.

Philippine peatlands

The Philippines have at least two areas with significant deposits of tropical peats. These are the Agusan Marsh and the Leyte Sab-a Basin. The Caimpugan peatland in the Agusan Marsh in Mindanao is one of the most ecologically significant wetlands in the Philippines. Nevertheless, the Caimpugan peatland (with its most important carbon store being its peat soil), is under threat of clearance and conversion to agricultural land. Some areas are also affected by tourism development and disturbance. Land-clearance is widespread in the peripheral peatland zones where there is tall-pole forest. The Caimpugan peatland which extends over 5,487 hectares has been estimated as holding around twenty-three million tonnes of carbon. The mire is a very significant contributor to carbon storage in the country but is currently under severe threat.

The Sab-a Basin peatland is in a basin close to the northern coast of Leyte with a total area of about 3,000 hectares, and of this, around 44% has been converted to agriculture. This leaves around 1,740 hectares still as peatland and made up of relatively small remnant areas of peat-swamp forest and sedge and grass peat-swamp. There are other smaller areas of mire also identified across the Philippines and the recent ASEAN Peatland Management Initiative (APMI) is at last beginning to highlight the importance of conservation and restoration here.

Oceania

The geographic region of Oceania includes Australasia, Melanesia, Micronesia, and formerly, (until the 1950s), Indonesia and the Philippines. These were removed from Oceania and added to Asia. This does make a significant difference to how we might view the peatlands of the region. However, in the diverse lands which make up Oceania, there are scattered peatlands of different forms and sizes – from peat-swamp forest to blanket mire. However, in almost all situations the mires are to an extent degraded by grazing of domestic stock, peat-mining, drainage with associated bushfires, and various other human-related impacts. One of the associated effects has been a marked increase in wildfires in countries such as Australia where hydrological functions have been compromises in landscapes which already tend towards drought and heat.

The mires of the Australian Alps

The sphagnum peatlands of Australia were discussed by Whinam et al. (2003) and the peat forests by Adam (1987, 1992).

The limited areas of fen and sphagnum bog in Australia are found in zones such as the Australian Alps but here they have declined dramatically over the past century and a half. Part of the problem is due to the activities of feral animals and the impacts of exotic plants. Desiccation and high temperatures have led to wildfires over most of the alpine, sub-alpine, and montane bogs and fens of the region. A single year of fires destroyed over 15% of the functional mires.

Work began in 2003 to restore mire systems and to restore the hydrological functions to these degraded ecosystems. Novel approaches to reinstating the vulnerable sphagnum sites included the use of shading cloth to protect them from intense ultraviolet light, and straw-bale dams to hold back water. This project restored about 400 hectares of damaged mire on 130 sites and this amounted to about 10% of the fire-damaged resource. A degree of recovery occurred in a period of from fifteen months to six years, but it was recognized that a return to a fully functional mire system will take much longer to achieve.

The Americas

The American mires range from the vast northern blanket and raised sphagnum bogs, to the extensive tropical peat-swamp forests of Central and South America. In the north there are extensive polygon mires in the arctic and sub-arctic zones, and in more southerly zones of North America there are raised bogs and peat-swamp forests, with extensive blanket mires and fens along both east and west coasts. Blanket mires also reappear at it the southerly zones of South America. However, in Central and South America, many of the tropical swamp-forest areas are under threat of drainage and burning, and across the USA and Canada peatlands have been drained, cut, and converted. Industrial-scale peat-mining and drainage to convert land to farming have destroyed huge areas of peatland habitat. Across the central zone of the Americas, sites are lost to urban development, to agriculture, and sometimes to tourism. The Amazonian forests are being drained, felled, and burned, and in Patagonia and the Andes, overgrazing by livestock is a serious threat.

Canadian peat industry

There are now some attempts to restore and reinstate drained and cut-over mires after industrial exploitation. The Canadian horticultural peat industry has been working to develop what they consider to be effective restoration techniques. Part of this is described as 'the moss-layer transfer technique' which involves the active reintroduction of peat bog plants and the improvement of the micro-environment combined with re-wetting. Given suitable climatic conditions the projects have achieved good coverage of bog vegetation dominated by sphagnum mosses within a few years. With ditch-blocking the water table rises and hydrological functions begin to return. However, the regenerated mire does exhibit fluctuations in water levels, and the project hopes this will stabilise and the peat-moss layer thickens over time. The researchers also hope that the annual carbon balance will be restored to near-natural state within about fifteen years. This remains to be seen.

The Florida Everglades

One of the major restoration projects in the USA has been in the Everglades National Park. However, despite some remarkable recovery since the troubles of the early twentieth century, the peatlands here still have serious issues. Much of the original landscape has been lost to intensive farming and what remains is still suffering from problems of contamination, water supply issues, and ecological challenges such as the spread of numerous non-native species.

The Great Dismal Swamp

The Great Dismal Swamp National Wildlife Refuge is the largest intact remnant of a vast peatland habitat which formerly extended over around half a million hectares of south-eastern Virginia and north-eastern North Carolina. Much of the surrounding peripheral land was drained and converted to farmland but the core site remained largely intractable. The National Wildlife Refuge covers around 50,000 hectares managed for nature conservation. In order to slow and reverse the impacts of extensive drainage there have been sluices constructed to hold back the waters. However, because of the drainage some parts of the reserve are now vulnerable to wildfires caused for example by lightning strikes. The intention is to recover the hydrological functioning of the area but with around 200 kilometres of drains the restoration is a long-term challenge.

A related restoration project is in the 50,000-hectare Pocosin Lakes National Wildlife Refuge which is in North Carolina, south of the Great Dismal Swamp. Again this area experienced the similar problems of drainage, conversion to agriculture, contamination, and burning; by blocking drains and through conservation management, it is hoped that the ecosystem functions and carbon storage will be restored.

A brief overview of North American peatland restoration

There are thorough accounts of the current situation with North American peatland restoration by, for example, Chimner et al. (2017) and González & Rochefort (2019). Human impacts on North American peatlands clearly changed with colonisation of the continent by Europeans. However, over many centuries, these landscapes were utilised for activities such as hunting, grazing, gathering fuel, and collecting foods such as berries or wildfowl. There is considerable literature from the mid-1800s onwards on the commercial possibilities for peat-mining. Indeed, the major impact of human exploitation has been that of twentieth-century peat-mining, initially for fuel, and then for horticulture. This was extensive throughout the peatlands but especially in eastern Canada, Minnesota, and Michigan. With the growing realisation that natural regeneration in these heavily peat-extracted bogs is very limited, such sites became the focus for North American peatland restoration projects. Such remediation attempts triggered awareness that novel techniques were necessary (Figure 14.3).

With peatland degradation increasing through commercial and industrial use together with drainage for agriculture and forestry and associated with mineral ore and oil extraction, the need for a more holistic approach also emerged. Activities such as mining and oil extraction brought with them massive impacts of associated infrastructures (i.e. access roads, seismic lines, powerlines, pipelines, and oil well stations). Consequently, awareness of the significant costs to society of peatland degradation in terms of lost ecosystem services. In this context, peatland restoration which is ecologically and economically effective becomes a priority.

FIGURE 14.3 November 1948 and a dried-out bog is on fire with peat-fire smog spreading far and wide near Milwaukee.

Furthermore, the close connections between peatlands and global carbon cycles mean that governments, regulators, and corporate citizens increasingly demand sound ecological restoration.

From the above drivers, North American peatland restoration began in the 1990s on post-peat-mining sites with sphagnum transfers. Restoration methods aim to establish vegetation, hydrology, and associated ecological processes with approaches that dovetail with the demands of diverse site histories and basic differences in mire type (e.g. fens, swamps, and bogs). Since fenlands are controlled by groundwater and surface-flow restoring their hydrology is often more complex than for bogs. There are therefore some big differences between bog and fen restoration. Fens, especially in western North America, are often dominated by sedges (Cyperaceae) and these may be reintroduced using live plugs from nearby donor sites, seed collected from the same simply scattered at the receiver site, bought or cultivated seedlings from locally sourced seed, or plants grown on geotextile mats which arrive as rolled turf.

Similarly, whilst degraded peatlands may be invaded by woody species, some European and North American sites are naturally dominated by trees and in the USA are termed 'swamps'. Reinstating a functioning ecology to these sites incurs specific problems because of the slow growth rates of some of the tree species and heavy browsing pressure from wild herbivores like deer; it may be necessary to provide protection from grazing for many years. Drainage of such swamps will have led to tree and shrub species favouring drier conditions and in many cases they will have been replanted for commercial forestry too. Rehabilitating these sites may necessitate a series of key steps:

1. Clearance of post-disturbance tree cover;
2. Restoration of a fully functional hydrology and re-wetting;
3. Replanting the desired tree species to move the system back to a more natural condition.

However, trees are susceptible to changes in hydrology and may not establish if the site is too wet and with periods of waterlogging. In order to address these issues it may be necessary to manipulate microtopography to aid the creation of suitable microhabitats for establishment.

Browsing and grazing are issues for peatland restoration around the world and frequently the problems relate to domestic stock. However, in North America, the issue is often grazing and browsing by wild ungulates, specifically deer. Overgrazed fenland may be restored by combined passive and active techniques; by simply removing the animals from the fen, in the first instance. However, even with animals removed, successful recovery may depend on regeneration from a seed bank and that may be variable. If such recovery does not occur, then a more active intervention may be needed with species reintroductions. Deciding what to introduce may raise issues and the process may be informed by palaeo-ecological studies or perhaps an analysis of seed banks. Nevertheless, there may be a degree of subjectivity in terms of the chosen reference point to which restoration is targeted, and for a heavily grazed site, it may not be possible to return to its prior 'natural' condition. Large-scale restoration may necessitate re-establishment of a diversity of plant species along with sphagnum mosses, including other mosses, sedges, grasses, ericaceous and non-ericaceous shrubs, and trees too.

In North America, the arguments for effective peatland restoration have grown to take in matters such as potential legal requirements, wetland mitigation and carbon banking, climate mitigation, and water quality and water supply. Furthermore, they are increasingly embedded in ideas of responsible ecosystem management for both industry and wider society.

It seems that North American peatland restoration has developed from basic trial and error to empirically based scientific approaches supported by sound practitioner skills. It has moved from small site-specific experiments to landscape-scale restoration to reconstruct hydrological connectivity and ecological function. Along with these changes the process has evolved from interested individual stakeholders (often academics and conservationists) to being owned by multiple stakeholders (including private, local, and regional governmental agencies, NGOs, businesses, and more). Nevertheless, despite the considerable progress made there are still critical gaps in research, and long-term recovery

FIGURE 14.4 Thorne Moors National Nature Reserve being re-wetted as an ambitious restoration project in England.

to hydrological and ecological functions remain aspirational. Furthermore, the irrecoverable damage to archaeology and palaeo-archaeology by destructive peat-mining must still be recognised. It is now possible to reinstate a degree of function to these degraded landscapes, but this does not restore them to their prior status and condition (Figure 14.4).

15

FENSCAPE AND PEAT BOG

A future nature

Future peatlands and peatscapes

As is so often the case, the positive steps as described earlier are to be welcomed and celebrated but they are really rather too little too late. We are very late in acting to halt the catastrophic losses not only in the tropics but even in our backyard. It is encouraging to witness the new initiatives, but they must now continue to grow over the decades and in some cases, it really will be centuries before these sites really become functioning peatlands again. Fads and fashions can have no place in this future peatscape vision because above all, it requires long-term application and dedication. In terms of peatlands past, our society has wrought the greatest destruction to nature that one could possibly conceive, and in some places this still goes on. We are now beginning a long process of reconstruction and rebirth (Figure 15.1).

For success, these new peatscapes must engage local people and be embedded into local economies. Through this they can help combat the undoubtedly serious environmental challenges that confront society today and in the future. It is these peatlands and their hinterlands that will alleviate floods and droughts, help climate-proof our regions, soak-up carbon from the atmosphere, enhance biodiversity, and even kick-start rural tourism. We can now show that by helping mop-up excess atmospheric carbon peatlands will mitigate or buffer the worst excesses of climate change. These are inherently landscape-scale features and this will help nature conservation and, in particular support biodiversity responses to massive environmental change. Indeed, for species to survive major changes they need joined-up landscapes with suitable habitats and habitat corridors. Along with major woodlands and forests, peatlands are the obvious way to deliver this.

FIGURE 15.1 Sphagnum moss close-up at Thorne Moors in England. The bog-mosses are an indicator of recovery and are being monitored by a network of specialists coordinated by the *Thorne and Hatfield Moors Conservation Forum*.

The projects described are just a small selection of the wide range of excellent initiatives currently on-going. Nevertheless, good as they are, they are still not enough. We really have little choice and as Sir John Lawton wrote about Britain, we need to deliver conservation projects which are bigger, better, and more joined. With rising sea levels and increasingly extreme storm events acting across the world, we need to re-envision our engagement with Nature. The major threats are widespread but will affect coastal zones in particular, and also, river valleys downstream from extensive upland catchments. These are landscapes zones where peatlands can make a significant difference. The new peatscapes must be a key part of a process of human adaptation for survival, a long-term vision where we accept and plan for environmental change. The alternatives may be cheaper as we continue to plan and act merely for the immediate and the short-term. However, this is like the Dutch boy with his finger plugging the hole in the dyke. It may work for a little while but is not sustainable.

The intention of this short volume has been to introduce peatlands, their ecology and functioning, and their history and heritage. Stemming from this introduction, the idea was to address the causes and consequences of peatland decline,

FIGURE 15.2 Explaining restoration work on the Solway Mosses in north-west England and a long-term project to recover an internationally important resource.

of their loss, degradation, and transformation, often to intensive agriculture or to agri-forestry such as palm-oil plantations. It is from this desperately low point in peatland history, perhaps in the latter stages of the twentieth century, that the wider importance of peatland remediation for ecosystem services and climate-change mitigation has really taken root. Yet despite this, the corporate and individual hypocrisy of peatland destruction continues almost unabated in gardens and garden centres around Britain, throughout Europe and globally. I liken this to demolishing England's iconic York Minster and using the stone to build a parish church (or even a garden rockery). Peat bogs and fens are 'biodiversity cathedrals' and yet we demolish them to create our back-garden flower borders...... (Figure 15.2).

Priorities for global peatlands based on those stated by Joosten et al. (2012) are given below:

1. To identify the location, extent, and condition of all peatlands;
2. To improve the assessment of greenhouse gas emissions from peatlands;
3. To conserve all reasonably intact peatlands;
4. To prevent further degradation of already degraded peatlands to include:

    a restoring degraded peatlands by re-wetting, reforestation (in for example tropical peat-swamps and some temperate peatlands), conservation, and in some cases, paludiculture

  b targeting financial resources to peatland conservation, restoration, and improved management generally
  c stimulating and applying existing and developing climate-related financing mechanisms to peatland conservation and restoration

5. To engage and involve local communities at the earliest possible stage and support communities to address issues like opportunity costs for new approaches, and reliance on unsustainable exploitation of peatlands;
6. To ensure that greenhouse gas criteria are included in credible certification and subsidy schemes for products derived from drained peatlands. These should be integrated into national procurement policies;
7. Share between countries and communities, experience and expertise on peatland conservation, restoration, and improved management approaches.

With these basic principles in mind, we can carry on the works begun in recent decades to halt further losses and to begin a process of rebuilding. Publications from, for instance, the IUCN peatland programmes cited earlier join with a rich literature on peat, peatlands, mires, and fens which have developed over the last few decades. Furthermore, the scientific accounts have now been joined by volumes on the practicalities, such as '*Conserving Bogs: The Management Handbook*' by Brooks & Stoneman (1997). Increasingly too, we also have comprehensive accounts of the mires and peatlands resources around the world as illustrated by Joosten et al. (2017). There are still gaps in global knowledge and particularly in the awareness of the importance of cultural landscapes and histories in the peatlands story. Indeed, I argue elsewhere that understanding our history is the key to knowing where and why we are as we are today. This knowledge of our present condition and past history is I suggest, essential in guiding future attempts at recovery, restoration, and conservation. Taking the example of Holland and the loss of almost all its peatlands by the 1970s, it is hugely informative to understand the landscape and economic history of the region. It was close-by that the Romans marvelled at local people who burnt 'mud' for heating for example. By medieval times the Dutch economic boom, the so-called 'Golden Age', was powered by easy and cheap energy from peat fuel. Essentially, this is the story of the destruction of Holland's peat bogs. The details of what happened were described by several scholars such as De Vries (1974), De Zeeuw (1978), and TeBrake (1985), and I wonder how many of those involved in modern-day peatland restoration have read any of these. The latter stages of landscape transformation were the extensive drainage and reclamation of polders and associated peatlands for agriculture and horticulture during the nineteenth and twentieth centuries. Furthermore, the environmental footprint of the Dutch peat-cutters and drainage engineers extended well beyond Holland as the skills, knowledge, and enthusiasm for 'improvement' spread internationally. The point is, that to understand the modern challenges we really need to have a sound knowledge of environmental history. This need can be compounded as a problem by the lack of effective cross-disciplinary exchange of research and knowledge. This in part at least helps to explain the relative scarcity of any meaningful input on heritage and cultural matters to most peatland restoration projects.

However, the key challenge today is in taking forwards the knowledge and enthusiasm we have for peat bogs, fens and their restoration to apply them in practice, to exchange skills and knowledge, and to engage a far wider audience than has been done so far.

BIBLIOGRAPHY

Adam, I.H. (ed.) (1979) *Papers on Peter May Land Surveyor 1749–1793.* T. and A. Constable, Edinburgh.

Adam, P. (1987) *New South Wales Rainforests: The Nomination for the World Heritage List.* National Park and Wildlife Service of New South Wales, Sydney.

Adam, P. (1992) *Australian Rainforests.* Oxford University Press, Oxford.

Alloway, R. (1869) *Peat Coal versus Pit Coal.* Reprinted by the British Library, London, 2011.

Anderson, T. (2006) *Life on the Levels; Voices from a Working World.* Birlinn Ltd, Edinburgh.

Anon. (1994) *Report of the Working Group on Peat Extraction and Related Matters.* The Department of the Environment, HMSO, Ruislip.

Anon. (1998) *Biomass Energy: Data, Analysis and Trends.* International Energy Agency, European Commission Research Directorate, United Nations Environment Programme, OECD, Paris.

Anon. (2017) *Blanket Bog: Outcomes Approach – Land Management Guidance.* Produce by Moors for the Future for the Uplands Management Group, Edale, Derbyshire.

Anon. (2018) *UK Peatland Strategy 2018–2040.* IUCN UK Peatland Programme, Edinburgh.

Ardron, P.A. (1999) *Peat Cutting in Upland Britain, with Special Reference to the Peak District.* Unpublished PhD Thesis, University of Sheffield, Sheffield.

Ardron, P.A., Rotherham, I.D., & Gilbert, O.L. (1996) The influence of peat-cutting on upland landscapes: case studies from the South Pennines. In: Landscapes – Perception, Recognition and Management: reconciling the impossible? Proceeding of the Landscape Conservation Forum Conference, 2–4 April, 1996, Sheffield. *Landscape Archaeology and Ecology*, **2**, 56.

Ardron, P.A., Rotherham, I.D., & Gilbert, O.L. (1997) Factors determining contemporary upland landscapes........a re-evaluation of the importance of peat-cutting and associated drainage, and the implications for mire restoration and remediation. In: *Blanket Mire Degradation. Causes, Consequences and Challenges.* Proceedings of the British Ecological Society Conference in Manchester, 1997. British Ecological Society and the Macaulay Land Use Research Institute, Aberdeen, 38–41.

Ardron, P.A., Rotherham, I.D., & Gilbert, O.L. (1998) Peat-cutting and upland landscapes: case-studies from the South Pennines. In: Landscapes.....Perception, Recognition and Management: reconciling the impossible? Proceedings of the conference held in Sheffield, UK, 2–4 April, 1996. *Landscape Archaeology and Ecology*, **3**, 65–69.

Bain, C.G., Bonn, A., Stoneman, R., Chapman, S., Coupar, A., Evans, M., Gearey, B., Howat, M., Joosten, H., Keenleyside, C., Labadz, J., Lindsay, R., Littlewood, N., Lunt, P., Miller, C.J., Moxey, A., Orr, H., Reed, M., Smith, P., Swales, V., Thompson, D.B.A., Thompson, P.S., Van de Noort, R., Wilson, J.D. & Worrall, F. (2011) IUCN UK Commission of Inquiry on Peatlands. IUCN UK Peatland Programme, Edinburgh.

Baird, A.J., Holden, J., & Chapman, P. (2009) A Literature Review of Evidence on Emissions of Methane in Peatlands. Report, UK Government Department of Environment Fisheries and Rural Affairs (Defra) Project SP0574, University of Leeds, Leeds, UK.

Balfour-Browne, F. (1926) The aquatic Coleoptera of the Wicken Fen area, Cambridgeshire. In: Gardiner, J.S. (ed.) *The Natural History of Wicken Fen. Part III*. Bowes & Bowes, Cambridge, 201–214.

Bambalov, N.N. (1996) Peatlands and peat resources of Belarus. In: Lappalainen, E. (ed.) *Global Peat Resources*. International Peat Society and Geological Survey of Finland, Jyskä, 57–60.

Bather, D.M., & Miller, F.A. (1991) *Peatland Utilisation in the British Isles*. Centre for Agricultural Strategy, Paper 21, University of Reading, Reading.

Berry, A.Q., Gale, F., Daniels, J.L., & Allmark, W. (eds.) (1996) *Fenn's and Whixall Mosses*. Clwyd Archaeology Service, Mold.

Berthelot, Ch. (1941) La Tourbe. *La Nature*, **3066**, 35–41.

Berthelot, Ch. (1942) Organisation et Avenir de la Tourbière Française. *L'Illustration*, **5187**, 101–103.

Bevis, T.A. (1983) *Strangers in the Fens: Huguenot/Walloon Communities at Thorney, Parson Drove, Guyhirn and Sandtoft and Some Adherents*. Privately published by T.A. Bevis, March, Cambridgeshire.

Bingham, R.K. (1987) *The Chronicles of Milnthorpe*. Cicerone Press, Milnthorpe.

Blizard, J. (1917) *The Value of Peat Fuel for the Generation of Steam*. Government Printing Bureau, Ottowa.

Bloom, A. (1944) *The Farm in the Fen*. Faber and Faber, London.

Bonn, A., Allott, T., Hubacek, K., & Stewart, J. (eds.) (2009) *Drivers of Environmental Change in Uplands*. Routledge, London.

Bonn, A., Holden, J., Parnell, M., Worrall, F., Chapman, P.J., Evans, C.D., Termansen, M., Beharry-Borg, N., Acreman, M.C., Rowe, E., Emmett, B., & Tsuchiya, A. (2010) Ecosystem services of peat – Phase 1. DEFRA Project code: SP0572.

Bonn, A., Reed, M.S., Evans, C.D., Joosten, H., Bain, C., Farmer, J., Emmer, I., Couwenberg, J., Moxey, A., Artz, R., Tanneberger, F., Unger, M., Smyth, M.A., & Birnie, D. (2014) Investing in nature: developing ecosystem service markets for peatland restoration. *Ecosystem Services*, **9**, September 2014, 54–65.

Bonn, A., Allott, T., Evans, M., Joosten, H., & Stoneman, R. (eds.) (2016) *Peatland Restoration and Ecosystem Services*. Cambridge University Press, Cambridge.

Bragg, O.M., Hulme, P.D., Ingram, H.A.P., & Robertson, R.A. (eds.) (1992) *Peatland Ecosystems and Man: An Impact Assessment*. British Ecological Society and International Peat Society, published by the Department of Biological Sciences, University of Dundee, Dundee.

Bray, L.S. (2015) *The Past and the Peat: Archaeology and Peatland Restoration on Exmoor*. Exmoor National Park, Dulverton.

Brooks, S., & Stoneman, R. (1997) *Conserving Bogs. The Management Handbook*. The Stationery Office, Edinburgh.

Buckland, P.C. (1979) *Thorne Moors: A Palaeoecological Study of a Bronze Age Site; A Contribution to the History of the British Insect Fauna.* Department of Geography Occasional Publication No. 8, University of Birmingham, Birmingham.

Buckland, P.C., & Dinnin, M.H. (1997) The rise and fall of a wetland habitat: recent palaeoecological research on Thorne and Hatfield Moors. *Thorne & Hatfield Moors Papers*, **4**, 1–18.

Burton, R. (1996) The peat resources of Great Britain (Scotland, England, Wales and Isle of Man). In: Lappalainen, E. (ed.) *Global Peat Resources.* International Peat Society and Geological Survey of Finland, Jyskä, 79–86.

Burton, R.G.O., & Hodgson, J.M. (1987) *Lowland Peat in England and Wales.* Soil Survey of England and Wales Special Survey No. 15, Harpenden.

Carroll, A.R. (2015) *Geofuels: Energy and the Earth.* Cambridge University Press, Cambridge.

Charman, D. (2002) *Peatlands and Environmental Change.* John Wiley & Sons, London.

Chimner, R.A., Cooper, D.J., Wurster, F.C., & Rochefort, L. (2017) An overview of peatland restoration in North America: where are we after 25 years? *Restoration Ecology*, **25** (2), 283–292.

Clay, G.D., Dixon, S., Evans, M.G., Rowson, J.G., & Worrall, F. (2012) Carbon dioxide fluxes and DOC concentrations of eroding blanket peat gullies. *Earth Surface Processes and Landforms*, **37** (5), 562–571.

Clymo, R.S. (1983) Peat. In: Gore, A.J.P. (ed.) *Ecosystems of the World 4A. Mires: Swamps, Bog, Fen and Moor.* Elsevier, Amsterdam, 159–224.

Clymo, R.S. (1992) Models of peat growth. *Suo*, **43**,127–136.

Collingwood, W.G. (1912) *Elizabethan Keswick. Extracts from the Original Account Books, 1564–1577, of the German Miners in the Archives of Augsburg.* The Cumberland and Westmorland Antiquarian Society, Kendal.

Colomer, F., & Lordier, C. (1903) *Combustibles industriels; houille, pétrole, lignite, tourbe, bois, charbon de bois, agglomérés, coke.* Vve Ch. Dunod, Paris.

Couch, G. (1993) *Fuel Peat – World Resources and Utilisation.* IEA Coal Research, London.

Cris, R., Buckmaster, S., Bain, C., & Bonn, A. (eds.) (2011) *UK Peatland Restoration: Demonstrating Success.* IUCN UK National Committee Peatland Programme, Edinburgh.

Cris, R., Buckmaster, S., Bain, C., & Reed, M. (eds.) (2014) *Global Peatland Restoration Demonstrating SUCCESS.* IUCN UK National Committee Peatland Programme, Edinburgh.

Cross, J.R. (1989) *Peatlands: Wastelands or Heritage? An Introduction to Bogs and Fens.* Wildlife Service Ireland, Stationery Office, Dublin.

Cundhill, P., & Whittington, G. (1992) The historical impact of man on the peat areas of fife. In: Bragg, O.M., Hulme, P.D., Ingram, H.A.P., & Robertson, R.A. (eds.) *Peatland Ecosystems and Man: An Impact Assessment.* British Ecological Society and International Peat Society, published by the Department of Biological Sciences, University of Dundee, Dundee, 392–395.

Davies, D. (1785) Cases of labourers in husbandry stated and considered. Privately published, Barkham, Berkshire, England (Quoted in: Page, W.H., & Ditchfield, P.H. (eds.) *Victoria County History: A History of the County of Berkshire*, **Volume 3**, 1923, 238–241, Institute of Historical Research, London).

Day, A. (1985) *Turf Village: Peat Diggers of Wicken.* Cambridgeshire Libraries & Information Service, Cambs.

De Vries, J. (1974) *The Dutch Rural Economy in the Golden Age, 1500–1700.* Yale University Press, New Haven and London.

De Zeeuw, J.W. (1978) Peat and the Dutch Golden Age – the historical meaning of energy-attainability. *A.A.G. Bijdragen*, **21**, 3–31.

Denyer, S. (1991) *Traditional Buildings & Life in the Lake District*. Victor Gollancz, London.

Donald, M.B. (1955) *Elizabethan Copper: The History of the Company of Mines Royal 1568–1605*. Pergammon Press, London.

Dreschel, T.W., Hohner, S., Aich, S., & McVoy, C.W. (2018) Peat soils of the Everglades of Florida, USA. In: Topcuoğlu, B., & Turan, M. (eds.) *Peat*. InTech, London, 29–45.

Dunn, H. (2012) *Accounting for Environmental Impacts: Supplementary Green Book Guidance*. HM Treasury / Defra, London.

Elkington, J. (1797) *An Account of the Most Approved Mode of Draining Land According to the System Practiced by Mr. Joseph Elkington: With an Appendix, Containing Hints for the Farther Improvement of Bogs and Other Marshy Ground, After Draining: Together with Observations on Hollow and Surface Draining in General*. Drawn up for consideration of the Board of Agriculture and Internal Improvement, by John Johnstone, London.

Elkington, T., Dayton, N., Jackson, D.L., & Strachan, I.M. (2001) *National Vegetation Classification: Filed Guide to Mires and Heaths*. JNCC, Peterborough.

Evans, E.E. (1942) *Irish Heritage: The Landscape, the People and their Work*. W. Tempest, Dungalgan Pres, Dundalk.

Evans, M., & Warburton, J. (2007) *Geomorphology of Upland Peat*. Blackwell Publishing Ltd, Oxford.

Evelyn, J. (1729) *Silva*. (1979 reprint of the fifth edition), Stobart and Son, London.

Feehan, J., & O'Donovan, G. (1996) *The Bogs of Ireland: An Introduction to the Natural, Cultural and Industrial Heritage of Irish Peatlands*. The Environmental Institute, University College, Dublin.

Fell, A. (1908) *The Early Iron Industry of Furness and District*. Hume Kitchin, Ulverston.

Fenton, A. (1986) *The Shape of the Past 2. Essays in Scottish Ethnology*. John Donald Publishers Ltd, Edinburgh.

Fernández-Garcia, J.M., & Pérez, F.J. (eds.) (2018) *Inventory, Value and Restoration of Peatlands and Mires. Recent Contributions*. HAZI Foundation, Bizkaia.

Fitzrandolph, H.E., & Hay, M.D. (1926) *The Rural Industries of England & Wales. A Survey made on behalf of the Agricultural Economics Research Institute Oxford. Volume II: Osier-growing and Basketry and Some Rural Factories*. Clarendon Press, Oxford.

Friday, L. (ed.) (1997) *Wicken Fen: The Making of a Nature Reserve*. Harley Books, Colchester.

Gailey, A., & Fenton, A. (eds.) (1970) *The Spade in Northern and Atlantic Europe*. Ulster Folk Museum and Institute of Irish Studies, Queen's University, Belfast.

Gambles, R. (1997) *The Story of the Lakeland Dales*. Phillimore & Co. Ltd, Chichester.

Gimingham, C.H. (1972) *Ecology of Heathland*. Chapman and Hall, London.

Godwin, Sir H. (1981) *The Archives of the Peat Bogs*. Cambridge University Press, Cambridge.

Godwin, H., & Conway, V.M. (1939) The ecology of a raised bog near Tregaron, Cardiganshire. *Journal of Ecology*, **27**, 313–359.

González, E., & Rochefort, L. (2019) Declaring success in Sphagnum peatland restoration: identifying outcomes from readily measurable vegetation descriptors. *Mires and Peat*, **24** (19), doi: 10.19189/MaP.2017.OMB.305.

Goodwillie, R. (1984) *European Peatlands*. European Committee for Conservation of Nature and Natural Resources, Nature and Environment Series, 19, Council of Europe, Strasbourg.

Gorham, E. (1953) Some early ideas concerning the nature, origin and development of peat lands. *Journal of Ecology*, **41** (2), 257–274.

Gorham, E. (1957) The development of peat lands. *The Quarterly Review of Biology*, **32** (2), 145–166.

Gorham, E. (1991) Northern peatlands: role in the carbon cycle and probable responses to climatic warming. *Ecological Applications*, **1** (2), 182–195.

Gorham, E., Lehman, C., Dyke, A., Janssens, J., & Dyke, L. (2007) Temporal and spatial aspects of peatland initiation following deglaciation in North America. *Quarterly Science Reviews*, **26**, 300–311.

Gorham, E., Lehman, C., Dyke, A., Clymo, D., & Janssens, J. (2012) Long-term carbon sequestration in North American peatlands. *Quarterly Science Reviews*, **58**, 77–82.

Granlund, E. (1932) De svenska högmossarnas geologi. *Sver. Geol. unders. Arsbok.*, **26** (1), 1–193.

Grant, I.F. (1961) *Highland Folk Ways*. Routledge & Kegan Paul, London.

Graves, A.R., & Morris, J. (2013) Restoration of Fenland Peatland under Climate Change. Report to the Adaptation Sub-Committee of the Committee on Climate Change. Cranfield University, Bedford.

Greeves, T.A.P. (1981) The archaeological potential of the Devon tin industry. CBA Research Report, **40**, *Medieval Industry*. Crossley, D.W. (ed.), Council for British Archaeology, London, 85–95.

Grunwald, M. (2006) *The Swamp. The Everglades, Florida and the Politics of Paradise*. Simon & Schuster, New York.

Gumbricht, T., Roman-Cuesta, R.M., Verchot, L., Herold, M., Wittmann, F., Householder, E., Herold, N., & Murdiyarso, D. (2017) An expert system model for mapping tropical wetlands and peatlands reveals South America as the largest contributor. *Global Change Biology*, doi: 10.1111/gcb.13689.

Haanel, E. (1918) *Peat as a Source of fuel*. Commission of Conservation, Canada, Ottawa.

Haines-Young, R., & Potschin, M. (2008) *England's Terrestrial Ecosystem Services and the Rationale for an Ecosystem Approach*. Overview Report, Defra, London.

Hall, D., Wells, C.E., & Huckerby, E. (1995) *The Wetlands of Greater Manchester. North West Wetlands Survey 2*. Lancaster Imprints, Lancaster.

Hamilton-Jenkins, A.K. (1932) *Cornwall and its People*. Dent and Sons Ltd., London.

Hammond, R.F. (1979) *The Peatlands of Ireland*. Soil Survey Bulletin No. 35, An Foras Talúntais, Dublin.

Handley, C., & Rotherham, I.D. (2011) *Mosses & Cloughs. Moor Memories in the Holme Valley Area*. Wildtrack Publishing, Sheffield.

Harrison, M.E., & Rieley, J.O. (2018) Tropical peatland biodiversity and conservation in Southeast Asia. *Mires and Peat*, **22** (00), 1–5.

Hartley, M., & Ingilby, J. (1968) *Life and Tradition in the Yorkshire Dales*. Dent and Sons, London.

Hartley, M., & Ingilby, J. (1972) *Life and Tradition in the Moorlands of North-East Yorkshire*. Smith Settle, Otley.

Heinemeyer, A., Croft, S., Garnett, M.H., Gloor, M., Holden, J., Lomas, M.R., Ineson, P. (2010) The MILLENNIA peat cohort model, predicting past, present and future soil carbon budgets and fluxes under changing climates in peatlands. *Climate Research*, **45**, 207– 226.

Heinemeyer, A., Asena, Q., Burn, W.L., & Jones, A.L. (2018) Peatland carbon stocks and burn history: blanket bog peat core evidence highlights charcoal impacts on peat physical properties and long-term carbon storage. *GEO: Geography and Environment*, e00063. doi: 10.1002/geo2.63.

Hodgkinson, D., Huckerby, E., Middleton, R., & Wells, C.E. (1995) *The Lowland Wetlands of Cumbria. North West Wetlands Survey 6*. Lancaster Imprints, Lancaster.

Hogan, D.V., & Maltby, E. (2005) The potential for carbon sequestration in wetlands of the Humberhead Levels. Technical report by the Royal Holloway Institute for Environmental Research, Royal Holloway University of London, Surrey.

Hohner, S.M., & Dreschel, T.W. (2015) Everglades peats: using historical and recent data to estimate predrainage and current volumes, masses and carbon contents. *Mires and Peat*, **16**, (2015), 1–15, Article 1.

Holden, J., Chapman, P.J., & Labadz, J.C. (2004) Artificial drainage of peatlands: hydrological and hydrochemical process and wetland restoration. *Progress in Physical Geography*, **28** (1), 95–123.

Holden, J., et al. (2007), Environmental change in moorland landscapes. *Earth Sci. Rev.*, **82**, 75–100.

Holman, I.P. (2009) *An Estimate of Peat Reserves and Loss in the East Anglian Fens Commissioned by the RSPB*. Department of Natural Resources, Cranfield University, Cranfield.

Howard-Davis, C., Stocks, C., & Innes, J. (1988) *Peat and the Past. A Survey and Assessment of the Prehistory of the Lowland Wetlands of North-west England*. English Heritage & Lancaster University, Lancaster.

Humphries, C.J., & Shaughnessy, E. (1987) *Gorse*. Shire Publications Ltd., Aylesbury.

Immirzi, C.P., & Maltby, E., with Clymo, R.S. (1992) *The Global Status of Peatlands and their Role in Carbon Cycling. A report for Friends of the Earth prepared by The Wetland Ecosystems Research Group, Department of Geography, University of Exeter*. University of Exeter, Exeter.

Ingram, H.A.P. (1967) Problems of hydrology and plant distribution in mires. *Journal of Ecology*, **55**, 711–724.

Ingram, H.A.P. (1982) Size and shape in raised mire ecosystems; a geophysical model. *Nature*, **297**, 300–303.

Ingram, H.A.P. (1983) Hydrology. In: Gore, A.J.P. (ed.) *Ecosystems of the World 4A, Mires: Swamp, Bog, Fen and Moor*. Elsevier, Oxford, 67–158.

Ingram, H.A.P., & Bragg, O.M. (1984) *The Diptotelmic Mire: Some Hydrological Consequences Reviewed*. In: Proceedings of the Seventh International Peat Congress, Dublin, 220–234.

Jauhiainen, J., Page, S.E., & Vasander, H. (2016) Greenhouse gas dynamics in degraded and restored tropical peatlands. *Mires and Peat*, **17** (06), 1–12.

Jenkins, J.G. (1965) *Traditional Country Craftsmen*. Routledge & Kegan Paul, London.

Jenkins, J.G. (ed.) (1969) *Studies in Folk Life. Essays in Honour of Iorwerth C. Peate*. Routledge & Kegan Paul, London.

Johansen, A. (1996) The extent and use of peatlands in Norway. In: Lappalainen, E. (ed.) (1996) *Global Peat Resources*. International Peat Society and Geological Survey of Finland, Jyskä, 113–117.

Joosten, H., Tapio-Bistrom, M., & Tol, S. (eds.) (2012) *Peatlands – Guidance for Climate Change Mitigation by Conservation, Rehabilitation, and Sustainable Use*. Second edition. Mitigation of Climate Change in Agriculture Series 5, Food and Agriculture Organisation of the United Nations; on-line at www.fao.org/climatechange/micca/peat/en/.

Joosten, H., Tanneberger, F., & Moen, A. (eds.) (2017) *Mires and Peatlands of Europe*. E. Scheitzerbart'sche Verlagsbuchhandlung, Stuttgart.

Kerr, W.A. (1905) *Peat and its Products*. Begg, Kennedy & Elder, Glasgow.

Kivinen, E. (1980) *Proposal for General Classification Virgin*. In: *The Proceedings of the 6th International Peat Congress*, 17–23 August, Duluth, Minnesota, 47.

Kivinen, E., & Pakarinen, P. (1980) *Peatland Areas and the Proportion of Virgin Peatlands in Different Countries*. In: *The Proceedings of the 6th International Peat Congress*, 17–23 August, Duluth, Minnesota, 52–54.

Kivinen, E., & Pakarinen, P. (1981) Geographical distribution of peat resources and major peatland complex types in the world. *Ann. Acad. Sci. Fennicae (Geol. Geog.) III*, **132**, 1–28.

Knox, S.A. (1985) *The Making of the Shetland Landscape*. John Donald Publishers Ltd, Edinburgh.

Kumar, H. (2002) *Encyclopaedia of Energy Resources and Their Management*. Volume 1. Ivy Publishing House, Delhi.

Kutsch, W.L., Bahn, M., & Heinemeyer, A. (2010) *Soil Carbon Dynamics: An Integrated Methodology*. Cambridge University Press, Cambridge.

Labadz, J.C. (1988) Runoff and sediment production in blanket peat moorland: studies in the southern Pennines. Unpublished PhD thesis. Huddersfield Polytechnic, Huddersfield.

Labadz, J.C., Burt, T.P., & Potter, A.W.R. (1991) Sediment yield and delivery in the blanket peat moorlands of the Southern Pennines. *Earth Surface Processes and Landforms*, **16**, 255–271.

Lambert, J.M., Jennings, J.N., Smith, C.T., Green, C., & Hutchinson, J.N. (1961) *The Making of the Broads. A Reconsideration of their Origin in the Light of New Evidence*. John Murray Ltd., London.

Lappalainen, E. (ed.) (1996a) *Global Peat Resources*. International Peat Society and Geological Survey of Finland, Jyskä.

Lappalainen, E. (1996b) General review on world peatland and peat resources. In: Lappalainen, E. (ed.) *Global Peat Resources*. International Peat Society and Geological Survey of Finland, Jyskä, 53–56.

Lindsay, R. (1992) *Conservation of Peat Bogs*. Nature Conservancy Council, Peterborough.

Lindsay, R. (1995) *Bogs: The Ecology, Classification and Conservation of Ombrotrophic Mires*. Scottish Natural Heritage, Battleby.

Lindsay, R. (2010) *Peatbogs and Carbon: A Critical Synthesis to Inform Policy Development in Oceanic Peat Bog Conservation and Restoration in the Context of Climate Change*. University of London, London.

Lowenthal, D. (1985) *The Past is a Foreign Country*. Cambridge University Press, Cambridge.

MAFF (1988) Agricultural Land Classification of England and Wales. *Revised Guidelines and Criteria for Grading the Quality of Agricultural Land*. MAFF, London.

Malby, E., & Proctor, M.C.F. (1996) Peatlands: their nature and role in the biosphere. In: Lappalainen, E. (ed.) *Global Peat Resources*. International Peat Society and Geological Survey of Finland, Jyskä, 11–19.

Maltby, E. (1997) Peatlands: the science case for conservation and sound management. In: Parkyn, L., Stoneman, R.E., & Ingram, H.A.P. (eds.) *Conserving Peatlands*. CAB International, Wallingford, 121–131.

Mann, S. (1993) *Cattle Cotton and Commerce. Life around Gatehouse of Fleet in 1792. A Reprint of the First Statistical Account of Scotland for the Parishes of Anworth and Girthon 1792*. Beechwood Publishing, Cambridge.

Marshall, J.D., & Davies-Shiel, M. (1971) *The Lake District at Work Past and Present*. David & Charles, Newton Abbot.

Martin-Ortega, J., Glenk, K., Byg, A., & Okumah, M. (2017) *Public Views and Values of Peatland Restoration in Scotland: Results of a Quantitative Study*. The James Hutton Institute and University of Leeds, Leeds.

Mawby, F.J. (1995) Effects of damming peat cuttings on Glasson Moss and Wedholme flow, two Lowland Raised Bogs in North-West England. In: Wheeler, B.D., Shaw, S.C., Fojt, W.J., & Robertson, R.A. (eds.) *Restoration of Temperate Wetlands*. John Wiley and Sons, Chichester, 349–358.

McVoy, C.W., Said, W.P., Obeysekera, J., VanArman, J.A., & Dreschel, T.W. (2011) *Landscapes and Hydrology of the Predrainage Everglades*. University Press of Florida, Gainesville.

Middleton, R., Wells, C.E., & Huckerby, E. (1995) *The Wetlands of North Lancashire. North West Wetlands Survey 3*. Lancaster Imprints, Lancaster.

Middleton, R., Tooley, M.J., & Innes, J.B. (2013) *The Wetlands of South West Lancashire: North West Wetlands Survey 7*. Lancaster Imprints, Lancaster.

Millennium Ecosystem Assessment (2005) *Ecosystems and Human Well-Being: Wetlands and Water – Synthesis*. World Resources Institute, Washington.

Milne, R., & Brown, T.A. (1997) Carbon in the vegetation and soils of Great Britain. *Journal of Environmental Management*, **49**, 413–433.

Mitchell, W.R. (2005) *Around Morecambe Bay*. Phillimore & Co. Ltd, Chichester.

Moen, A. (1999) *National Atlas of Norway: Vegetation*. Norwegian Mapping Authority, Hønefoss.

Moore, P.J., & Bellamy, D. (1974) *Peatlands*. Paul Elek Scientific Books Ltd, London.

Natlandsmyr, B., & Hjelle, K.L. (2016) A long-term vegetation dynamics and land-use history: providing a baseline for conservation strategies in protected *Alnus glutinosa* swamp woodlands. *Forest Ecology and Management*, **372**, 78–92.

Orru, M. (1996) Peat resources of Estonia. In: Lappalainen, E. (ed.) *Global Peat Resources*. International Peat Society and Geological Survey of Finland, Jyskä, 144–148.

Page, S.E., & Rieley, J.O. (1998) Tropical peatlands: a review of their natural resource functions, with particular reference to Southeast Asia. *International Peat Journal*, **8**, 95–106.

Page, S.E., Siegert, F., Rieley, J.O., Boehm, H.D.V., Jaya, A., & Limin, S. (2002) The amount of carbon released from peat and forest fires in Indonesia during 1997. *Nature*, **420**, 61–65.

Page, S.E., Rieley, J.O., & Wüst, R. (2006) Chapter 7: Lowland tropical peatlands of Southeast Asia. In: Martini, I.P., Cortizas, A.M., & Chesworth, W. (eds.) *Peatlands: Evolution and Records of Environmental and Climate*, Elsevier B.V., Amsterdam, 145–172.

Page, S., Hosciło, A., Wösten, H., Jauhiainen, J., Silvius, M., Rieley, J., Ritzema, H., Tansey, K., Graham, L., Vasander, H., & Limin, S. (2009) Restoration ecology of lowland tropical peatlands in Southeast Asia: current knowledge and future research directions. *Ecosystems*, **12**, 888–985.

Page, S.E., Rieley, J.O., & Banks, C.J. (2011) Global and regional importance of the tropical peatland carbon pool. *Global Change Biology*, **17** (2), 798–818.

Parkyn, L., Stoneman, R.E., & Ingram, H.A.P. (eds.) (1997) *Conserving Peatlands*. CAB International, Wallingford.

Parry, M.L. (1977) *Mapping Moorland Change: A Framework for Land-Use Decisions in the Peak District*. Peak District National Park, Bakewell.

Paulus, C. (1927) *Unpublished pages relating to the Manor and Parish of Ecclesall including the Enclosures of the Common and Waste Lands there*. Sheffield.

Pearce, F. (2017) Can we find the world's remaining peatlands in time to save them? *Yale Environment 360*, April 4, 2017.

Peate, I.C. (1972) *Tradition & Folk Life: A Welsh View*. Faber & Faber Ltd, London.

Peckham, S.F. (ed.) (1874) *Peat for Domestic Fuel. The Geological and Natural History Survey of Minnesota*. Tribune Publishing Company, Minneapolis.

Peter, D. (1994) *In and Around Silverdale. The Story of a North Lancashire Village*. Barry Ayre, Carnforth.

Pilkington, M. et al. (2015) *Natural Flood Management: An Appraisal of Current Status*. Moors for the Future Partnership, Edale, Derbyshire.

Pinchbeck, I. (1930) *Women Workers and the Industrial Revolution 1750–1850*. George Routledge & Sons, London.

Pinches, S. (2007) From common rights to cold charity: enclosure and poor allotments in the eighteenth and nineteenth centuries. In: Borsay, A., & Shapely, P. (eds.) *Medicine Charity and Mutual Aid.* Ashgate Publishing Limited, Aldershot, 35–54.

Price, J.S., Heathwaite, A.L., & Baird, A.J. (2003) Hydrological processes in abandoned and restored peatlands: an overview of management approaches. *Wetlands, Ecology and Management*, **11**, 65–83.

Proctor, M.C.F. (1995a) Hydrochemistry of the raised bogs and fens at Malham Tarn National Nature Reserve, Yorkshire, UK. In: Hughes, J.M.R., & Heathwaite, A.L. (eds.) *Hydrology and Hydrochemistry of British Wetlands.* Wiley and Sons, Chichester, 273–289.

Proctor, M.C.F. (1995b) The ombrogenous bog environment. In: Wheeler, B.D., Shaw, S.C., Fojt, W.J., & Robertson, R.A. (eds.) *Restoration of Temperate Wetlands.* John Wiley & Sons, Chichester, 287–304.

Proctor, M.C.F., & Maltby, E. (1998) Relations between acid atmospheric deposition and the surface pH of some ombrotrophic bogs in Britain. *Journal of Ecology*, **86**, 329–340.

Rackham, O. (1986) *The History of the Countryside.* Dent, London.

Raistrick, A., & Jennings, B. (1965) *A History of Lead Mining in the Pennines.* Longmans, London.

Robertson, R.A., & Jowsey, P.C. (1968) *Peat Resources and Development in the United Kingdom.* In: *The Proceedings of the 3rd International Peat Congress, 18–23 August, Quebec*, 13–14.

Robroek, B.J.M., Jassey, V.E.J., Payne, R.J., Martí, M., Bragazza, L., Bleeker, A., Buttler, A., Caporn, S.J.M., Dise, N.B., Kattge, J., Zając, K., Svensson, B.H., van Ruijven, J., & Verhoeven, T.A. (2017) Taxonomic and functional turnover are decoupled in European peat bogs. *Nature Communications*, doi: 10.1038/s41467-017-01350-5.

Rogers, J.E.T. (1902) *A History of Agriculture and Prices in England.* Volumes 1–8. Clarendon Press, Oxford.

Rollinson, W. (1996) *The Lake District Life and Traditions.* Weidenfeld & Nicolson, London.

Rotherham, I.D. (1995) Urban heathlands.....their conservation, restoration and reclamation. *Landscape Contamination and Reclamation*, **3** (2), 99–100.

Rotherham, I.D. (1999a) Peat cutters and their landscapes: fundamental change in a fragile environment. *Landscape Archaeology and Ecology*, **4**, 28–51.

Rotherham, I.D. (ed.) (1999b) *Peatland Ecology and Archaeology: Management of a Cultural Landscape.* Wildtrack Publishing, Sheffield.

Rotherham, I.D. (2005) Fuel and landscape – exploitation, environment, crisis and continuum. *Landscape Archaeology and Ecology*, **5**, 65–81.

Rotherham, I.D. (2008a) Landscape, water and history. *Practical Ecology and Conservation*, **7**, 138–152.

Rotherham, I.D. (2008b) Lessons from the past – a case study of how upland land-use has influenced the environmental resource. *Aspects of Applied Biology*, **85**, *Shaping a vision for the uplands*, 85–91.

Rotherham, I.D. (2009) *Peat and Peat Cutting.* Shire Publications, Oxford.

Rotherham, I.D. (2010) *Yorkshire's Forgotten Fenlands.* Pen & Sword Books Limited, Barnsley.

Rotherham, I.D. (2011a) Habitat fragmentation and isolation in relict urban heaths – the ecological consequences and future potential. In: Rotherham, I.D., & Bradley, J. (eds.) (2009) *Lowland Heaths: Ecology, History, Restoration and Management.* Wildtrack Publishing, Sheffield, 106–115.

Rotherham, I.D. (2011b) Cultural severance in landscapes and the causes and consequences for lowland heaths. In: Rotherham, I.D., & Bradley, J. (eds.) *Lowland Heaths: Ecology, History, Restoration and Management.* Wildtrack Publishing, Sheffield, 130–143.

Rotherham, I.D. (2013a) War & Peat: exploring interactions between people, human conflict, peatlands, and ecology. In: Rotherham, I.D., & Handley, C. (eds.) (2013) *War & Peat*. Wildtrack Publishing, Sheffield, 7–44.

Rotherham, I.D. (2013b) *The Lost Fens: England's Greatest Ecological Disaster*. The History Press, Stroud.

Rotherham, I.D. (2013c) A fear of nature – images & perceptions of heath, moor, bog & fen in England. In: Joanaz de Melo, C., Queiroz, A.I., da Silveira, L.E., & Rotherham, I.D. (eds.) *Between the Atlantic and the Mediterranean. Responses to Climate and Weather Conditions throughout History*. Wildtrack Publishing, Sheffield, 131–162.

Rotherham, I.D. (2014a) *Eco-History: A Short History of Conservation and Biodiversity*. The White Horse Press, Cambridge.

Rotherham, I.D. (2014b) The call of the wild. Perceptions, history people & ecology in the emerging paradigms of wilding. *ECOS*, **35** (1), 35–43.

Rotherham, I.D. (2014c) *Spas and Spa Visiting*. Shire Publications, Oxford.

Rotherham, I.D. (2015) *The Rise and Fall of Countryside Management*. Routledge, London.

Rotherham I.D. (2020) History and heritage in the bog – examples from Cumbria and the surrounding areas. In Rotherham, I.D. (ed.) *History & Heritage of the Bogs & Peatlands of Cumbria and Surrounding Areas*. Wildtrack Publishing, Sheffield.

Rotherham, I.D., Egan, D., & Ardron, P.A. (2004) Fuel economy and the uplands: the effects of peat and turf utilisation on upland landscapes. *Society for Landscape Studies*, 2, 99–109.

Rotherham, I.D., & Egan, D. (2005) The economics of fuel wood, charcoal and coal: an interpretation of coppice management of British woodlands. In: Agnoletti, M., Armiero, M., Barca, S., & Corona, G. (eds.) *History and Sustainability*. European Society for Environmental History, 100–104.

Rotherham, I.D., Griffiths, T., & Handley, C. (2013) Sphagnum: the healing harvest. In: Rotherham, I.D., & Handley, C. (eds.) *War & Peat*. Wildtrack Publishing, Sheffield, 201–219.

Rotherham, I.D., & Handley, C. (eds.) (2013) *War & Peat*. Wildtrack Publishing, Sheffield.

Rotherham, I.D., & Handley, C. (eds.) (2020) *In The Bog*. Wildtrack Publishing, Sheffield.

Rotherham, I.D., & McCallam, D. (2008) Peat bogs, marshes and fen as disputed landscapes in late eighteenth-century France and England. In: Lyle, L., & McCallam, D. (eds.) *Histoires de la Terre: Earth Sciences and French Culture 1740–1940*. Rodopi B.V., Amsterdam & New York, 75–90.

Roucoux, K.H., Lawson, I.T., Jones, T.D., Baker, T.R., Coronado, E.N.H., Gosling, W.D., & Lähteenojac, O. (2013) Vegetation development in an Amazonian peatland. *Palaeogeography, Palaeoclimatology, Palaeoecology*, **374**, 15 March 2013, 242–255.

Rydin, H., & Jeglum, J. (2006) *The Biology of Peatlands*. Oxford University Press, Oxford.

Sanchez, P. (2016) Estimation of the carbon content of the Humberhead Levels wetlands over time: the case of Thorne Moors. Unpublished MSc dissertation, Sheffield Hallam University, Sheffield.

Shaler, N.S. (1895) *Origin, Distribution, and Commercial Value of Peat Deposits*. Government Printing Office, Washington.

Siever, R. (1980) *Energy and Environment*. Scientific American, W.H. Freeman and Company, San Fransisco.

Smiles, S. (1904) *Lives of the Engineers*. John Murray, London.

Smith, D. (ed.) (March 1951) *The Third Statistical Account of Scotland, XX1X. The County of Kincardine*. Scottish Academic Press, Edinburgh.

Smout, C. (2000) *Nature Contested – Environmental History in Scotland and Northern England Since 1600*. Edinburgh University Press, Edinburgh.

Snore, A. (1996) The peat resources of Latvia. In: Lappalainen, E. (ed.) *Global Peat Resources*. International Peat Society and Geological Survey of Finland, Jyskä, 101–105.

Soper, E.K., & Osbon, C.C. (1922) *The Occurrence and Uses of Peat in the United States*. Government Printing Office, Washington.

Suffling, E.R. (1885) *The Land of the Broads: A Practical Guide for Yachtsmen, Anglers, Tourists, and Others Pleasure-Seekers on the Broads and Rivers of Norfolk and Suffolk*. Benjamin Perry, London.

Sutherland, P., & Nicolson, A. (1986) *Wetland: Life in the Somerset Levels*. Michael Joseph, London.

Tallis, J.H., Meade, R. & Hulme, P.D. (eds.) (1997) *Blanket mire degradation. Causes, consequences and challenges*. Proceedings of the British Ecological Society Conference in Manchester, 1997. British Ecological Society and the Macaulay Land Use Research Institute, Aberdeen.

Tallis J.H. (1998) The southern Pennine experience: an overview of blanket mire degradation. In: Tallis, J.H., Meade, R., & Hulme, P.D. (eds.) *Blanket Mire Degradation*. Proceedings, British Ecological Society, Aberdeen, 7–15.

Tansley, A.G. (1939) *The British Islands and their Vegetation*. Cambridge University Press, Cambridge.

TeBrake, W.H. (1985) *Medieval Frontier. Culture and Ecology in Rinjland*. A. & M. University Press, Texas.

TEEB (2008) *The Economics of Ecosystems and Biodiversity: An Interim Report*. European Commission, Brussels.

TEEB (2009) *The Economics of Ecosystems and Biodiversity*, Climate Issues Update. 32p.

TEEB (2010a) *The Economics of Ecosystems and Biodiversity for Local and Regional Policy Makers*; on-line at www.TEEBweb.org.

TEEB (2010b) *The Economics of Ecosystems and Biodiversity: Ecological and Economic Foundations*. Edited by Pushpam Kumar. Earthscan, London.

TEEB (2011a) *The Economics of Ecosystems and Biodiversity in Business*. Edited by Joshua Bishop, Earthscan, London.

TEEB (2011b) *The Economics of Ecosystems and Biodiversity in National and International Policy Making*. Edited by Patrick ten Brink. Earthscan, London.

TEEB (2011c) *The Economics of Ecosystems and Biodiversity in Local and Regional Policy and Management*. Edited by Heidi Wittmer and Haripriya Gundimeda. Earthscan, London.

Thompson, F. (1968) *Harris and Lewis: Outer Hebrides*. David & Charles, Newton Abbot.

Turp, J.S. (1916) *Peat in 1915*. Government Printing Office, Washington.

Van De Noort, R., & Ellis, S. (eds.) (1997) *The Wetland Heritage of the Humberhead Levels*. University of Hull Press, Hull.

Van der Valk, A., Volin, J.C., & Wetzel, P. (2015) Predicted changes in interannual water-level fluctuations due to climate change and its implications for the vegetation of the Florida Everglades. *Environmental Management*, **55** (4), doi: 10.1007/s00267-014-0434-4.

Van der Zee, B. (2017) UK is 30–40 years away from 'eradication of soil fertility', warns Gove. *The Guardian*, Tuesday 24th October, 2017.

Vasander, H., Tuittila, E.S., Lode, E., Lundin, L., Ilomets, M., Sallantaus, T., Heikkilä, R., Pitkänen, M.L., & Laine, J. (February 2003) Status and restoration of peatlands in northern Europe. *Wetlands Ecology and Management*, **11** (1–2), 51–63.

Warde, P., & Williamson, T. (2014) Fuel supply and agriculture in post-medieval England. *Agricultural History Review*, **62**, 61–82.

Watson, R., & Albon, S. (2011a) *UK National Ecosystem Assessment: Understanding Nature's Value to Society. Technical Report.* UNEP-WCMC, Cambridge.

Watson, R., & Albon, S. (2011b) *UK National Ecosystem Assessment: Understanding Nature's Value to Society. Synthesis of the Key Findings.* UNEP-WCMC, Cambridge.

Webb, N. (1986) *Heathlands.* Collins, London.

Webb, N.R. (1998) The traditional management of European heathlands. *Journal of Applied Ecology*, **35**, 987–990.

Wellard, G. (1995) *200 Years of Frimley's History. The Story of the Frimley Fuel Allotments Charity and Pine Ridge Golf Centre.* Pine Ridge Golf Centre and Frimley Fuel Allotments Charity, Camberley, Surrey.

Wentworth-Day, J. (1954) *History of the Fens.* George Harrap & Co. Ltd., London.

Wheeler, B.D. (1980) Plant communities of rich-fen systems in England and Wales: II. Communities of Calcareous Mires. *Journal of Ecology*, **68**, (2), 405–420.

Wheeler, B.D., & Proctor, M.C.F. (2000) Ecological gradients, subdivisions and terminology of northwest European Mires – essay review. *Journal of Ecology*, **88**, 187–203.

Wheeler, B.H., & Shaw, S.C. (1995) *Restoration of Damaged Peatlands.* Department of the Environment, HMSO, London.

Whinam, J., Hope, G.S., Clarkson, B.R., Buxton, R.P., Alspach, P.A., & Adam, P. (2003) Sphagnum in peatlands of Australasia: their distribution, utilisation and management. *Wetlands Ecology and Management*, **11**, 37–49.

Wickman, F.E. (1951) The maximum height of raised bogs and a note on the motion of water in soligenous mires. *Geologiska Föreningens i Stockholm Förhandlingar*, **73**, 419.

Williams, M. (1970) *The Draining of the Somerset Levels.* Cambridge University Press, Cambridge.

Williamson, T., & Warde, P. (2013) Seeing the wood for the trees: the diversity of local fuel sources and the transition to a coal burning economy in England. Conference paper at the Rural History Conference 2013, Bern.

Withington, L. (ed.) (1899) *Elizabethan England.* Scott, London.

Worrall, F., Harriman, R., Evans, C.D., Watts, C., Adamson, J., Neal, C., Tipping, E., Burt, T.P., Grieve, I., Montieth, D., Naden, P.S., Nisbet, T., Reynolds, B., & Stevens, P. (2004) Trends in dissolved organic carbon in UK rivers and lakes. *Biogeochemistry*, **70** (3), 369–402.

Wright, L. (1964) *Home Fires Burning......The History of Domestic Heating and Cooking.* Routledge, London.

Young, A. (1772) *A six weeks tour through the southern counties of England and Wales: describing particularly: with descriptions and copper-plates, of such newly invented implements of husbandry as deserve to be generally known: interspersed with accounts of the seats of the nobility and gentry, and other objects worthy of notice / by the author of The farmer's letters.* Printed for Strahan, W., Nicoll, W., Cadell, T., Collins, B., & Balfour, J., London.

Young, A. (1804) *General View of the Agriculture of the County of Norfolk.* The Board of Agriculture, London.

INDEX

For Product Safety Concerns and Information please contact our EU representative GPSR@taylorandfrancis.com Taylor & Francis Verlag GmbH, Kaufingerstraße 24, 80331 München, Germany

Batch number: 10777124

Printed by Printforce, the Netherlands